D1481905

THE FRONTIERS COLLECTION

THE FRONTIERS COLLECTION

Series Editors:
D. Dragoman M. Dragoman A.C. Elitzur M.P. Silverman J. Tuszynski H.D. Zeh

The books in this collection are devoted to challenging and open problems at the forefront of modern physics and related disciplines, including philosophical debates. In contrast to typical research monographs, however, they strive to present their topics in a manner accessible also to scientifically literate non-specialists wishing to gain insight into the deeper implications and fascinating questions involved. Taken as a whole, the series reflects the need for a fundamental and interdisciplinary approach to modern science. It is intended to encourage scientists in all areas to ponder over important and perhaps controversial issues beyond their own speciality. Extending from quantum physics and relativity to entropy, time and consciousness – the Frontiers Collection will inspire readers to push back the frontiers of their own knowledge.

Information and Its Role in Nature
By J.G. Roederer

Relativity and the Nature of Spacetime
By V. Petkov

Quo Vadis Quantum Mechanics?
Edited by A. C. Elitzur, S. Dolev, N. Kolenda

Life – As a Matter of Fat
The Emerging Science of Lipidomics
By O.G. Mouritsen

Quantum–Classical Analogies
By D. Dragoman and M. Dragoman

Knowledge and the World
Challenges Beyond the Science Wars
Edited by M. Carrier, J. Roggenhofer, G. Küppers, P. Blanchard

Quantum–Classical Correspondence
By A.O. Bolivar

Mind, Matter and Quantum Mechanics
By H. Stapp

Quantum Mechanics and Gravity
By M. Sachs

J. G. Roederer

INFORMATION AND ITS ROLE IN NATURE

With 35 Figures

RECEIVED

NOV - 1 2006

MINNESOTA STATE UNIVERSARY LIBRARY
MANKATO, MN 56002-8419

Springer

Professor Dr. Juan G. Roederer
University of Alaska
Geographical Institute
Koyukuk Drive 903
Fairbanks, AK 99775-7320, USA

QH
507
.R62
2005

Series Editors:

Prof. Daniela Dragoman
University of Bucharest, Physics Faculty, Solid State Chair, PO Box MG-11,
76900 Bucharest, Romania   email: danieladragoman@yahoo.com

Prof. Mircea Dragoman
National Research and Development Institute in Microtechnology, PO Box 38-160,
023573 Bucharest, Romania   email: mircead@imt.ro

Prof. Avshalom C. Elitzur
Bar-Ilan University, Unit of Interdisciplinary Studies,
52900 Ramat-Gan, Israel   email: avshalom.elitzur@weizmann.ac.il

Prof. Mark P. Silverman
Department of Physics, Trinity College,
Hartford, CT 06106, USA   email: mark.silverman@trincoll.edu

Prof. Jack Tuszynski
University of Alberta, Department of Physics, Edmonton, AB,
T6G 2J1, Canada   email: jtus@phys.ualberta.ca

Prof. H. Dieter Zeh
University of Heidelberg, Institute of Theoretical Physics, Philosophenweg 19,
69120 Heidelberg, Germany   email: zeh@urz.uni-heidelberg.de

Cover figure: Detail from 'Zero of $+1/-1$ Polynomials' by J. Borwein and L. Jorgensen. Courtesy of J. Borwein

ISSN 1612-3018
ISBN-10 3-540-23075-0 Springer Berlin Heidelberg New York
ISBN-13 978-3-540-23075-5 Springer Berlin Heidelberg New York

Library of Congress Control Number: 2005924951

This work is subject to copyright. All rights are reserved, whether the whole or part of the material is
concerned, specifically the rights of translation, reprinting, reuse of illustrations, recitation, broadcasting,
reproduction on microfilm or in any other way, and storage in data banks. Duplication of this publication or
parts thereof is permitted only under the provisions of the German Copyright Law of September 9, 1965, in
its current version, and permission for use must always be obtained from Springer. Violations are liable to
prosecution under the German Copyright Law.

Springer is a part of Springer Science+Business Media.
springeronline.com

© Springer-Verlag Berlin Heidelberg 2005   Printed in Germany

The use of general descriptive names, registered names, trademarks, etc. in this publication does not imply,
even in the absence of a specific statement, that such names are exempt from the relevant protective laws and
regulations and therefore free for general use.

Typesetting by Stephen Lyle using a Springer TEX macro package
Final processing by LE-TEX Jelonek, Schmidt & Vöckler GbR, Leipzig
Cover design by KünkelLopka, Werbeagentur GmbH, Heidelberg

Printed on acid-free paper   SPIN: 10977170   57/3141/YL - 5 4 3 2 1 0

6258240

To my children

Ernesto, Irene, Silvia and Mario
... my greatest pride and joy

Foreword

According to some, the Age of Information was inaugurated half a century ago in the nineteen forties. However, we do not really know what information is. By "we" I mean everybody who is not satisfied with the trivial meaning of information as "what is in the paper today," nor with its definition as "the number of bits in a telegraph" and even less with the snappy "negentropy." But we do feel that the revival of the ancient term in a modern scientific discourse is timely and has already started a quiet revolution in our thinking about living matter, about brains and minds, a revolution perhaps leading to a reunification of Culture, for centuries tragically split between the Humanities and Science.

Who are the natural philosophers whose thinking is broad enough to encompass all the phenomena that are at the basis of the new synthesis? Erwin Schrödinger comes to mind with his astonishing book "What is life?", astonishing because it comes from one who had already been a revolutionary in the foundation of the new physics. Another one is Norbert Wiener whose mathematical-poetic Cybernetics was powerful enough to penetrate even the farthest reaches of futuristic science fiction (cybernauts in cyberspace, etc.). But also Juan Roederer comes to mind.

Here is one whom I have heard producing torrents of baroque music on a majestic pipe organ. One who learned his physics from the greatest masters of quantum theory including Werner Heisenberg but did not deem it beneath him to apply his art to the down-to-earth subject of geophysics (of the magnetosphere, to be sure). One who regaled us with a classic text on the physics and psychophysics of music. A theorist who does not shy away from the theory of his own self. One who did his homework very thoroughly in modern biology and brain science.

In our search for a proper place of "information" in a Theory of the World we have been barking up the wrong tree for centuries. Now we are beating around the bush from all sides. Reading Roederer I get the impression that he knows exactly where the prey is hiding.

Valentino Braitenberg
Director Emeritus
Max Planck Institute for Biological Cybernetics
Tübingen, Germany
January 2005

Preface

The roots of this book can be traced back to the early nineteen seventies. At that time I was teaching, as a departmental out-reach action, a course on musical acoustics at the University of Denver. Preparing material for my classes, I began to realize that the acoustical information-processing inside the head was as interesting a topic for a physicist as the physics of what happens outside, in the instrument and the air. As a consequence, the scope of my lectures was expanded to include the mechanisms of musical sound perception. This led me to some "extracurricular" research work on pitch processing, the organization of the "Workshops on Physical and Neuropsychological Foundations of Music" in Ossiach, Austria, and the first edition of my book "Physics and Psychophysics of Music". It did not take me long to become interested in far more general aspects of brain function, and in 1976 I organized a course called "Physics of the Brain" (a title chosen mainly to circumvent departmental turf conflicts). Stimulated by the teaching and the discussions with my students, I published an article in *Foundations of Physics,* [91] launching my first thoughts on information as the fundamental concept that distinguishes physical interactions from the biological ones – sort of central theme of the present book.

My directorship at the Geophysical Institute of the University of Alaska and, later, the chairmanship of the United States Arctic Research Commission prevented me from pursuing this "hobby" for many years. In 1997 I became Senior Adviser of the International Centre for Theoretical Physics (ICTP) in Trieste, Italy; my occasional duties there offered me the opportunity to participate and lecture at Julián Chela Flores' fascinating astrobiology and neurobiology summer schools and symposia. This put me back on track in the interdisciplinary "troika" of brain science, information theory and physics. Of substantial influence on my thinking were several publications, most notably B.-O. Küppers' book "Information and the Origin of Life" [64] and J. Bricmont's article "Science of Chaos or Chaos of Science?" [21], as well as enlightening discussions with Valentino Braitenberg at his castle (yes, it is a castle!) in Merano, Italy.

I am deeply indebted to Geophysical Institute Director Roger Smith and ICTP Director Katepalli Sreenivasan for their personal encouragement and institutional support of my work for this book. Without the help of the Geophysical Institute Digital Design Center and the competent work of Kirill Maurits who produced the illustrations, and without the diligent cooperation of the ICTP Library staff, particularly chief librarian Maria Fasanella, the preparation of the manuscript would not have been possible.

My special gratitude goes to Valentino Braitenberg of the Max Planck Institute for Biological Cybernetics in Tübingen, my son Mario Roederer of the National Institutes of Health in Bethesda, Maryland, GianCarlo Ghirardi of the ICTP and the University of Trieste, Daniel Bes of the University Favaloro in Buenos Aires, and Glenn Shaw of the University of Alaska Fairbanks, who have read drafts of the manuscript and provided invaluable comments, criticism and advice.

And without the infinite patience, tolerance and assistance of my wife Beatriz, this book would never have materialized.

Juan G. Roederer
Geophysical Institute, University of Alaska Fairbanks and
The Abdus Salam International Centre for Theoretical Physics, Trieste
http://www.gi.alaska.edu/~Roederer
February 2005

Contents

Introduction

Is Yet Another Book on Information Needed?

We live in the Information Age. Information is shaping human society. Great inventions facilitating the spread of information such as Gutenberg's movable printing type, radio-communications or the computer have brought about explosive, revolutionary developments. Information, whether good, accidentally wrong or deliberately false, whether educational, artistic, entertaining or erotic, has become a trillion dollar business. Information is encoded, transformed, censored, classified, securely preserved or destroyed. Information can lead nations to prosperity or into poverty, create and sustain life or destroy it. Information-processing power distinguishes us humans from our ancestor primates, animals from plants and bacteria from viruses. Information-processing machines are getting faster, better, cheaper and smaller. But the most complex, most sophisticated, somewhat slow yet most exquisite information-processing machine that has been in use in its present shape for tens of thousands of years, and will remain so for a long time, is the human brain. Our own self-consciousness, without which we would not be humans, involves an interplay in real time of information from the past (instincts and experience), from the present (state of the organism and environment), and about the future (desires and goals) – an interplay incomprehensively complex yet so totally coherent that it appears to us as "just one process": the awareness of our one-and-only self with a feeling of being in total, effortless control of it.

This very circumstance presents a big problem when it comes to understanding the concept of information. Because "Information is Us," we are so strongly biased that we have the greatest difficulty to detach ourselves from our own experience with information whenever we try to look at this concept scientifically and objectively. Like pornography, "we know it when we see it" – but we cannot easily define it! Working definitions of information of course have been formulated in recent decades, and we will discuss them here – but they refer mostly to particular applications, to limited classes of informational systems, to certain physical domains, or to specific aspects of the concept. When some years ago I asked an expert on information theory what information *really* is, he replied, somewhat in despair and without any further elaboration: "All of the above!"

So what is this powerful yet "ethereal" something that resides in CDs, books, sound waves, is acquired by our senses and controls our behavior, sits in the genome and directs the construction and performance of an organism? It is *not* the digital pits on the CD, the fonts in the books, the oscillations of air pressure, the configuration of synapses and distribution of neural activity in the brain, or the bases in the DNA molecule – they all *express* information, but they are not *the* information. Shuffle them around or change their order ever so slightly – and you may get nonsense, or destroy an intended function! On the other hand, information can take many forms and still mean the same – what counts in the end is what information *does,* not how it looks or sounds, how much it is, or what it is made of. Information has a *purpose,* and the purpose is, without exception, to cause some specific *change* somewhere, some time – a change that otherwise would not occur or would occur only by chance. Information may lay dormant for eons, but it is always intended to cause some specific change. How much information is there in a sheep? We may quote the number of bits necessary to list the bases of its DNA in correct order, and throw in an estimate of new synapses grown since its birth. But that number would just be some measure of the *amount* of information; it still would not give us anything remotely related to a sheep – it would have no connection with the purpose and potential effect of the information involved! In summary, information is a *dynamic* concept. Can we comprehend and describe this "ethereal thing" within the strict framework of physics?

Any force acting on a mass point causes a change of its velocity, and the rate of change is proportional to that force, as Newton's law tells us. The force is provided by some interaction mechanism (say, the expanding spring in a physics lab experiment, a gravitational field, a muscular effort or the electric field between two charged bodies), and there is a direct interchange between the energy of the mechanism and that of the mass point – a simple cause-and-effect relationship. However, when the cause for change is information, no such simple relationship exists: When I walk in a straight line through a corridor and see an exit sign pointing to the right, my motion also changes in a specific way – but in this case a very complex chain of interaction mechanisms and cause-and-effect relationships is at work. Note that the original triggering factor was a sign: Information is embedded in a particular *pattern* in space or time – it does not come in the form of energy, forces or fields, or anything material, although energy and/or matter are necessary to carry the information in question.

Information may be *created* (always for a purpose, like when the above mentioned exit sign was made and installed), *transmitted* through space and preserved or *stored* throughout time, but it also may be *extracted.* When I take a walk in the forest, I try not to run into any tree that stands in my way. Information about the tree is extracted by my visual system and my body reacts accordingly. Nobody, including the tree, has created the necessary information as a planned warning sign for humans. Rather, it is a pattern (the tree's silhouette, projected onto my retina) out of which information is extracted. The "purpose" in this example does not lie in the original pattern, but in the mechanism that allows me to perceive the pattern and react accordingly. And I will react accordingly only if that pattern has some *meaning* for me. Speaking of warning signs, consider those black circular eye-like dots on the wings of the Emperor moth. They are patterns which may have a clear message to birds: Stay away, I might be a cat or an owl! This is one of zillions of examples of purposeful information that *emerges* from a long process of "trial and error" in Darwinian evolution and is stored in the DNA molecules of each species. Random errors in DNA structure and their propagation to descendants, caused by quantum processes, perturbations from the chemical environment, radiation or some built-in transcription error mechanism, play the fundamental physical role in this process.

The above examples concerning information involve living beings. It is characteristic that even the most primitive natural mechanisms responding to information are complex, consisting of many interacting parts and involving many linked cause-and-effect relationships. What about nonliving things like computers, servomechanisms, particle detectors, robots, photographic cameras, artificial intelligence systems, etc.? These are all inanimate *artifacts*, devices planned and built by humans for a specific purpose. They may be simple or complex. Whatever information they create, extract, transmit or process is in response to some plan or program ultimately designed by a human brain. Such artifacts are informational systems of a special class, which as a rule will *not* be discussed explicitly in this book.

Information always has a *source* or *sender* (where the original pattern is located or generated) and a *recipient* (where the intended change is supposed to occur). It must be *transmitted* from one to the other. And for the specific change to occur, a specific physical mechanism must exist and be activated. We usually call this action *information processing.* Information can be *stored* and *reproduced,* either in the form of the original pattern, or of some *transformation* of it. And here comes a crucial point: For the transformation to embody the *same* information, it must somehow be able to lead to the *same* intended change in the recipient. It is the intended effect, what ultimately identifies information (but note that this does not yet define what information per se is!).

At this stage, it is important that you, the reader, agree with me that a fundamental property of information is that *the mere shape or pattern of something – not its field, forces or energy – can trigger a specific change in a recipient,* and do this consistently over and over again (of course, forces and energy are necessary in order to effect the change, but they are subservient to the purpose of the information in question). This has been called the pragmatic aspect of information [64]. It is important to emphasize again that the pattern alone or the material it is made of is *not* the information per se, although we are often tempted to think that way. Indeed, it is impossible to tell a priori if a given pattern contains information: For instance, one cannot tell by examining a complex organic molecule or a slice of brain tissue whether or not it possesses information (beyond that generated in our senses by its own visual appearance) – complexity and organization alone do not represent information [29]. As hinted above, information must not only have a purpose on part of the sender, it must have a meaning for the recipient in order to elicit the desired change.

Some typical questions to be addressed in this book are (not in this order): Is information reducible to the laws of physics and chemistry? Are information and complexity related concepts? Does the Universe, in its evolution, constantly generate new information? Or are information and information-processing exclusive attributes of *living* systems, related to the very definition of life? If that were the case (as this book posits), what happens with the physical meanings of entropy in statistical thermodynamics and wave function in quantum mechanics? What is the conceptual difference between classical and quantum information? How many distinct classes of information and information processing do exist in the biological world? How does information appear in Darwinian evolution? Does the human brain have unique properties or capabilities in terms of information processing? In what ways does information processing bring about human self-consciousness?

This book is divided into six closely intertwined chapters. The first chapter presents the basic concepts of classical information theory, casting it into a framework that will prepare the conceptual and mathematical grounds for the rest of the book, particularly the second chapter which deals with quantum information. Special emphasis is given to basic concepts like the "novelty value" of information, Shannon's entropy, and the "classical bit" as a prelude to the "quantum bit." The second chapter includes a brief introduction to quantum mechanics for readers who are not familiar with the basic tenets of this discipline. Since the principal focus of this book is to provide an insight into the concept of information *per se* wherever it appears, only the most fundamental aspects of classical and quantum information theory will be addressed in the first two chapters. Emphasis will be on the difference between the concepts of classical and quantum information, and on the counterintuitive behavior of quantum systems and its impact on the understanding of quantum computing. The third chapter represents the core of the book – and not just in the geometric sense! It dwells on the process of interaction as an "epistemological primitive," and posits the existence of two distinct classes of interactions, one of which is directly linked to, and defines, the concept of information in an objective way. The links of this "pragmatic" information with the more traditional but restricted concepts of algorithmic and statistical (Shannon) information are discussed. The fourth chapter, focusing on the role of information in biology, proceeds from "macroscopic" systems like neuron networks in the nervous system to the microscopic biomolecular picture. The main objective here is to show that information plays the *defining* role in life systems. In the fifth chapter we build on the preceding chapters and posit that information as such plays *no* active role in natural inanimate systems, whether classical or quantum – this concept enters physics *only* in connection with the observer, experimenter or thinker. Two fundamental concepts are taken as focal points in the discussion: entropy and the measurement process. Finally, since references to human brain function pervade the entire text, the last chapter serves as an introduction to recent findings on, and speculations about, the cognitive and affective functions of the brain. It attempts to show how difficult questions regarding mental processes such as animal consciousness, the formation of the concept of time, human thinking and self-consciousness, can be formulated in more objective terms by specifically focusing on information and information-processing.

My mission as university teacher during five decades has always been to help demystify as much as possible all that which physics and natural science in general is potentially able to demystify. Therefore I will take a reductionist, physicalist, "Copenhaguenist," biodeterministic and linguistic-deterministic stand. As a practicing space plasma physicist I am no expert in philosophy – which means that I have (thank heavens) no preconceived opinions in related matters. Philosophers have been asking questions during millennia about Nature, human beings and their place in it. Today, however, we must recognize that answers can only be found by following in the strictest way all tenets of the scientific method – logical analysis alone cannot lead to a quantitative understanding of the Universe. Surely, philosophers should continue pressing ahead with poignant questions; this will serve as a powerful stimulant for us "hard" scientists in the pursuit of answers! But careful: There will always remain some questions about the *why* of things that science cannot and should not address when such metaphysical questions deal with subjects not amenable, at least not yet, to objective measurement, experimentation and verification. Some will pop up explicitly or implicitly in the following pages; they indeed better be left to philosophers and theologians!

Since this book covers a vast spectrum of inter- and intradisciplinary topics, many subjects had to be treated only very superficially or left out altogether. For instance, there is nothing on technology, practical applications and details of laboratory experiments. Readers looking for descriptions of communications systems and data transmission, the workings of classical computers and the potential design of quantum computers, the mathematics and logic of information and probability, the genome, protein synthesis, cloning, behavioral aspects of brain function, information in the societal realm or historical notes will be disappointed. References are limited mostly to review articles in journals of more general availability; detailed literature sources can be found in the books cited. Unfortunately, many topics are still rather speculative and/or controversial – I apologize to the researchers working in the various intervening disciplinary areas for taking sides on issues where there is no consensus yet (or simply bypassing them), and for occasionally sacrificing parochial detail to improve ecumenical understanding.

The attempt to find a strictly objective and absolutely general definition of information is no trivial matter. In recent years more and more physicists, chemists and biologists have sought new definitions of information, more appropriate for the description of biological processes and which at the same time clarify the meaning of information in classical and quantum physics. Most of the existing articles are highly specialized and largely "monodisciplinary," and many books are conference proceedings or collections of chapters written by different authors. There are, of course, exceptions, but these are books written mainly for scientists already familiar with the new developments and they seldom present the matter from a truly interdisciplinary point of view. The present book expressly addresses the "middle class" – scientists across the disciplinary spectrum in the physical and life sciences, as well as university students at the graduate and upper undergraduate levels, who are interested in learning about a multidisciplinary subject that has not yet gained the broad attention it deserves. It requires knowledge of linear and matrix algebra, complex numbers, elementary calculus, and basic physics.

I am not aware of many single-authored books that take a fresh look at the concept of information and bind together its multiple roles in classical and quantum information theory, fundamental physics, cosmology, genetics, neural networks and brain function. This is, indeed, the goal I have set for myself in writing this book. Whether I have succeeded, only the reader can tell.

1 Elements of Classical Information Theory

1.1 Data, Information and Knowledge: The "Conventional Wisdom"

Scientists consider themselves experts in matters concerning information. Rightly so: The handling of information is the bread-and-butter for any scientific research endeavor, whether experimental or theoretical. In experimental research, data are acquired, analyzed and converted into information; information is converted into scientific knowledge, and scientific knowledge in turn poses new questions and demands for more data. This, in essence, is the merry-go-round of scientific research. What are a scientist's usual, intuitive, day-to-day views of the concepts of data, information and knowledge?

We usually think of the concept *data* as embodying sets of numbers that encode the values of some physical magnitude measured with a certain device under certain circumstances. And we usually think of the concept "information" as what is conveyed by statements that answer preformulated questions or define the outcome of expected alternatives. In terms of an expression of the amount of information everybody knows that the answer to a "yes or no question" represents one bit (short for binary unit) of information. In physics, the alternatives are often the possible states of a physical system, and information usually comes as a statement describing the result of a measurement. Data are meaningless without the knowledge of the device or paradigm used for their acquisition, the units, instrumental errors, codes and software used, and the particular circumstances of their acquisition – this is called the *metadata* pertaining to a given data base. Information is meaningless without knowledge of the questions or alternatives that it is supposed to answer or resolve, or without a description of the repertoire of possible states of a system that is being measured. While data and information can be handled by a computer, *knowledge* has to do exclusively with information gain by the human brain, represented in some very specific ways in its neural networks, and with the potential *use* of the gained information. In science, this use is mainly driven by the desire to make predictions about new situations, not yet seen or experienced, or to make "retrodictions" about an unknown past.

Data must be subjected to some mathematical algorithm in order to extract the information that provides answers to preformulated questions. Usually, in science we are dealing with questions about the properties or behavior of some simplified, idealized, approximate *model* of the system under consideration. The perhaps simplest case of information extraction in the case of a macroscopic system is that of calculating the average or expectation value $\langle x \rangle$ of N successive measurements x_i of a given physical magnitude which is supposed to have one unique value. The algorithm "average $\langle x \rangle = 1/N \sum_{i=1}^{N} x_i$"[†] yields a response to the question: What is the most probable value of the magnitude that we have measured? The x_i are the N data which have been converted into information ($\langle x \rangle$) about the physical magnitude. This does require a model: The idealization of the object being measured by assuming that a *unique* value of the magnitude in question does indeed exist and remain immutable during the course of the measurement (in the quantum domain this assumption is in general invalid), and that this value $\langle x \rangle$ is such that the algebraic sum of the errors $\varepsilon_i = (\langle x \rangle - x_i)$ is zero: $\sum \varepsilon_i = \sum (\langle x \rangle - x_i) = 0$ (or, equivalently, for which $\sum \varepsilon_i^2 = \min$). A second algorithm, the calculation of the standard deviation $\sigma = \left(\sum \varepsilon_i^2 / N \right)^{1/2} = \left(\langle x^2 \rangle - \langle x \rangle^2 \right)^{1/2}$ ($\langle x^2 \rangle$ is the average of the x_i^2), provides a measure of the quality or accuracy of the procedure used.[‡] Here the assumed model has to do with the statistical distribution of the measurement errors. For instance, two sets of data can have practically the same average value, yet differ greatly in their distribution about their average.

There are situations, however, in which the x_i do not represent multiple measurement values of the same constant magnitude, but are the result of individual measurements done on the elements of a *set* of many objects – for instance, the body weight of each individual in a group of persons. In that case, the "model" has to do with human biology and depends on gender, age, race and social conditions of the individuals measured; the value of $\langle x \rangle$ does not refer to the weight of *one* subject (there may be nobody in the group having that average weight), and σ does not represent any quality or accuracy of measurement but provides biological information about diversity in human development. In statistical mechanics the average of a string of data and corresponding standard deviation serve to establish a link between the microscopic and macroscopic variables of a body.

Finally, since a good part of natural science is concerned with finding out cause-and-effect relationships, another very fundamental example of information extraction is the determination of the correlation between two physical magnitudes x and y. The data are *pairs* x_i, y_i of simultaneous measurements, fed into an algorithm that furnishes the relevant information: the parameters of the functional relationship between the two magnitudes (pro-

---

[†] When the limits of a sum are obvious (e.g., the same as in a preceding expression), we shall omit them henceforth.

[‡] Note that for $N = 1$ (one data point) σ would be 0, which is absurd. For small samples, one really must use the *variance* $\nu = \left(\sum \varepsilon_i^2 / (N - 1) \right)^{1/2}$, which for $N = 0$ is indeterminate.

vided by some mathematical model), and the coefficient of correlation r, measuring the degree of confidence in the existence of a causal relationship. For a postulated linear relationship $y = ax + b$ (called linear regression), the expressions for the parameters are $a = \left(\langle xy \rangle - \langle x \rangle \langle y \rangle \right) / \left(\langle x^2 \rangle - \langle x \rangle^2 \right)$, $b = \left(\langle x \rangle^2 \langle y \rangle - \langle x \rangle \langle xy \rangle \right) / \left(\langle x^2 \rangle - \langle x \rangle^2 \right)$, and, for the correlation coefficient, $r = \left(\langle xy \rangle - \langle x \rangle \langle y \rangle \right) / \left[\left(\langle x^2 \rangle - \langle x \rangle^2 \right) \left(\langle y^2 \rangle - \langle y \rangle^2 \right) \right]^{1/2}$ (the coefficients a and b are derived from the condition that the sum of the squares of the "vertical" errors (ordinates) $\eta_i = y_i - (ax_i + b)$ be minimum). Interchanging x and y in these equations, we obtain the coefficients for the inverse linear regression $x = py + q$ (obtained by minimizing the sum of squares of "horizontal" errors (abscissae) $\xi_i = x_i - (py_i + q)$); the expression for the correlation coefficient r remains the same. The two regression lines $y = y(x)$ and $x = x(y)$ are *not* the same; both intersect at the point $\langle x \rangle$, $\langle y \rangle$ and the angle between them is a graphic measure of the degree of correlation (varying between zero for $r = 1$ and $90°$ for $r = 0$).

In general, information extraction algorithms are far more complicated. A remote sensing satellite image is nothing but a collection of data representing light emission intensities in a two-dimensional array of solid-angle pixels. Information is extracted from such data only when a given *pattern* is searched for with pattern-recognition software or when a human being is just looking at the picture and letting the brain recognize the pattern sought. Usually, this pattern is a particular feature of an idealized model of the physical system under study. The trace of an electrocardiogram is nothing but the graphic representation of a collection of data representing voltage signals picked up by some electrodes placed on the skin of a patient. Information is extracted only when certain patterns are searched for and recognized by the brain of the examiner or by some computer program (designed by a human brain). The patterns in question are features of an idealized model of the traumatized heart.

In the two above examples, the data appear in, or have been converted into, a form of sensorially detectable signals with the human cognitive apparatus – the brain – effecting the information extraction. What every scientist will recognize, but seldom take explicitly into consideration in his/her endeavor is that, ultimately, information extraction from any kind of data must *always* engage the human brain at some stage. If not in the actual process of information extraction, a brain will have been engaged at some stage in the formulation of the alternatives or questions to which the information to be extracted refers to, and also in the planning and construction of the instruments and experimental methods used. In science we may say that information only becomes information when it is recognized as such by a brain (more on this in later chapters) – data will remain data, whether we use it or not.

What is one person's information may well be another person's data. For instance, if we have M sets of N data each, all obtained under similar conditions, the average values $\langle x \rangle_k$ of each set can be considered data, and a "grand" average $\langle\langle x \rangle\rangle = \sum \langle x \rangle_k / M$ be considered new information. The standard deviation with which the individual averages $\langle x \rangle_k$ fluctuate around $\langle\langle x \rangle\rangle$ turns out to be approximately $\xi = \sqrt{\sum \varepsilon_i^2 / N(N-1)} \approx \sigma/\sqrt{N}$ for large N, the *standard deviation of the mean* (with the advantage that it can be calculated using data from only one of the sets in question). The content of information itself is often expressible in some quantitative form and thus can become data out of which information can be extracted at some higher level. One thus obtains the hierarchical chains of information extraction processes common to practically all research endeavors. An example is the conversion of raw data (often called "level I" data, e.g., [92]), such as the electric output pulses of a particle detector or the telemetry signals from a remote sensing satellite, to level II data which usually represent the values of a physical magnitude determined by some algorithm applied to level I data. A remote sensing image and an electrocardiogram trace are examples of level II data. Similarly, level III data are obtained by processing level II data (usually from multiple data suites) with the use of mathematical models so that information can be extracted on some global properties of the system under observation by identifying given patterns in the level II data. A weather map is a typical example of level III data – it becomes the input for a weather forecast. A most fundamental algorithm in all this is, indeed, that of *mapping:* the establishment of a one-to-one correspondence between clearly defined elements or features of two given sets.

So far we have discussed data and information. What about knowledge? This concept is a "hot potato," the discussion of which would drive us into philosophy and neurobiology. In the Introduction I declared philosophy off limits in this book, so this leaves us with the neurobiological aspects of knowledge. These will be treated in Chap. 6 with a pertinent introduction in Chap. 4.

For the theoretician, the limits between data, information and knowledge are blurred. Based on information gathered by the experimentalist on the correlation between measurable quantities – the "observables"[†] – the theoretician constructs mathematically tractable but necessarily approximate models of the systems under consideration, and formulates physical laws that allow quantitative prediction-making or the postdiction of their past evolution (see Chap. 5). Knowledge is what enables the theoretician to devise these models, to formulate the pertinent laws and to make the predictions or retrodictions (Chap. 5).

[†] Usually the term "observable" is reserved to designate measurable quantities in the quantum domain; for the classical equivalents the term "variables" is most frequently used.

1.2 Playing with an Idealized Pinball Machine

So far we have used the concept of information only as it is commonly understood in daily life (look up the rather naïve and often circular definitions given in dictionaries!) – we have not provided any scientifically rigorous definition of this concept. The classical theory of information (e.g., [22, 102–104]) quantifies certain aspects of information, but it does not propose any universal definition of information applicable to all sciences, nor does it deal explicitly with the concept of knowledge per se. It is not interested in the meaning conveyed by information, the purpose of sending it, the motivation to acquire it and the potential effect it may have on the recipient. Shannon's theory [102–104] is mainly focused on communications, control systems and computers; it defines a mathematical measure of the *amount* of information contained in a given message in a way that is independent of its meaning, purpose and the means used to generate it. It also provides a quantitative method to express and analyze the degradation of information during transmission, processing and storage. This theory is linked to statistical processes and works with quantitative expressions of the uncertainty about the outcome of possible alternatives and the amount of information expected to be received, once one such alternative has actually occurred. This is the reason why the term *statistical information* is used in this context. In general, in Shannon's theory the alternatives considered are messages drawn from a given finite, predetermined pool, such as dots and dashes, letters of the alphabet, sequences of numbers, words, addresses, etc., each message having a previously known probability to occur which may, or may not, depend on previous messages. Again, no consideration is given to the meaning and intention or purpose of the messages.

To emphasize the intimate link with statistical events, classical information theory usually begins with "tossing coins and throwing dice." We will work with an idealized standard "binary pinball machine," which emulates all basic features of a random coin toss but has better quantifiable components (Fig. 1.1). Initially, the ball is in the top bin. Once it is released, there are two available paths for the ball leading to two possible final states: one in which the ball is located in bin labeled 0 and another in which it is found in bin 1. We shall designate those possible states with the symbols $|0\rangle$ and $|1\rangle$, respectively (a notation borrowed from quantum mechanics – see next chapter). If the construction of the machine is perfectly symmetrical, repeating the process many times will reveal that each final state occurs 50% of the time. Note that under these conditions the operation of our standard pinball machine is totally equivalent to a coin toss; the reason we prefer to use this machine is that it has standard and controllable components that we will examine in detail and compare to other systems – including quantum systems. Our machine depicts two possible final states corresponding to the final position of the ball – we call these the *external* states of the system. We can also envision another machine with only one external state (just one

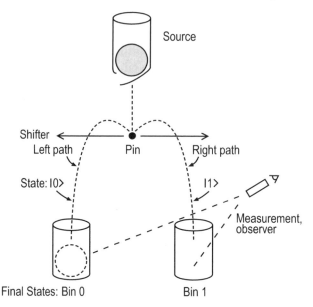

Fig. 1.1. Sketch of a "standard binary pinball machine" to be used in "gedanken-experiments" in this section. Initially, the ball is in the upper bin. After operation, the machine has two possible, mutually exclusive, final states: "ball in the left bin" or "ball in the right bin," designated (encoded) with the binary numbers 0 and 1, respectively. During the operation, the state is "ball on left path" or "ball on right path"; each one is coupled to the corresponding final state. An exactly centered position of the pin makes both states equally probable. Shifting the pin slightly to one side or the other will change the probabilities of occurrence of the final states

collecting bin) but drawing from a source containing a randomized mixture of balls of two different colors, representing different *internal* states.

Let us take a rather pedantic, piecemeal look into our machine and its operation. We can describe the final state, after the machine has been operated once, with words ("ball in left bin" or "ball having followed the left path," "red ball," "ball in right bin," etc.), with a symbol (as we did above), or with a number (the bin labels). If we use the binary number system (zeros and ones) as is done most frequently in information theory, we only need *one* binary digit, or *bit*, to specify the end state (that is why we called it a *binary* pinball machine in the first place!).

The trap door in the top bin, the pin, the running board and the gravitational field (coin tosses or pinball machines do not work on orbiting spacecraft!) represent the physical mechanism responsible for the achievement of the final state. The pin itself is the agent that determines the binary selection: To have a truly random outcome (a random choice of the final path), it is essential that the motion of the ball, once it has left the upper bin, be slightly randomized through small imperfections and stochastic perturbations (deterministic chaos – Sect. 5.2) that make the actual path followed after hitting the pin critically dependent on smallest changes of the incident direction – otherwise the final state would always be the same (similar randomization occurs when we flip a coin).

Finally, once the final state has been achieved, *we must look* into at least one of the two bins in order to know the outcome. We could also have a mechanical or electronic device registering the outcome; whichever way we do it, this process is called a *measurement,* and a human brain has to be involved at one time or another – making the observation or designing a device for that purpose and eventually looking at its record. Without a measurement there is no new information to be obtained! In view of the fact that we can express quantitatively what we have learned in the form of one binary digit, we say that this single measurement has provided us with 1 b *of new information.* An apparently trivial remark is the following: The act of measurement will in no way influence the result – we know for sure that the ball will follow either the right path or the left path, fall into the corresponding bin and stay there *whether we look or not;* the final state is defined in the interaction between the ball and the pin, and has nothing to do with any measurement process that follows. This remark, however, is not so trivial at all: It is generally *not* true for quantum systems!

We must emphasize that we are talking about amount of information, not its purpose or meaning: At this stage it is irrelevant whether knowledge of the outcome leads to reactions such as "So what?" or "I won the game!" Still, there is a crucial point: *Before* the measurement, we can assign a "value" to one of the two alternatives related to the likelihood or prior probability of its occurrence. Intuitively, the more probable a given outcome, the less should be its information value, and vice versa. Classical information theory defines an objective *information value* or *novelty value* for the outcome of each alternative, as we shall see below. Now, *after* having made the measurement we have acquired knowledge of the result: A transition has occurred in the cognitive state of our brain, from uncertainty to a state of certainty. Each state has a specific neural correlate (we shall discuss in detail what this means in Chap. 4 and Chap. 6), but Shannon's theory does not concern itself with the brain and any subjective aspects like purpose and meaning, or with any subjective value such as having won a game. It does, however, provide a mathematical expression for the *information gain* expected on the average before an alternative is resolved (a measurement is made), defined by the weighted average (with the respective probabilities as the weights) of the information values of each alternative. The more similar the probabilities of occurrence are (0.5 in our binary machine), the closer to 1 b should be the average information gain; the more dissimilar they are, which happens if one of the probabilities is close to 1 (certainty), the closer to 0 b it should be (because we knew ahead of time which alternative was most likely to come out!).

To show how these two measures concerning statistical information, also called *Shannon information,* are introduced, let us now tamper with our pinball machine (Fig. 1.1) and shift the pin a tiny amount to the left. There will still be two possible final states, but $|1\rangle$ will occur more frequently than $|0\rangle$. This is equivalent to tossing a loaded coin. If the pin shift exceeds a certain limit, the *only* possible state will be $|1\rangle$. If we operate the machine N times (with $N \to \infty$) under exactly the same conditions, and if N_0 and N_1 are the number of occurrences of $|0\rangle$ and $|1\rangle$, respectively, the ratios $p_0 = N_0/N$ and $p_1 = N_1/N$ are defined as the *probabilities* of occurrence, or prior probabilities, of states $|0\rangle$ and $|1\rangle$ ($p_0 + p_1 = 1$). Note that to have valid statistics, we must assume that the machine does not change in any way during use. In the case of an exaggerated shift of the pin, one of the probabilities will be $= 1$, the other 0; in the perfectly symmetric case, $p_0 = p_1 = 0.5$ – both end states are *equiprobable*. In the case of one of the p being 1, we say that the device has been *set* to a certain outcome (or that the final state has been set); there is no randomness in the outcome, but the act of setting could be the result of a random process at some prior level. For the time being, however, we will only consider random processes generated *inside* the pinball machine without external intervention.

Having *prepared* our machine, i.e., set the pin, determined the probabilities of the end states by operating it many times and found that they are *not* equal, we can no longer expect that one possible result will have the same novelty value as the other, nor that the average information gain would be 1 b. Indeed, if for instance $p_1 = 1$ (hence $p_0 = 0$), we would gain *no information at all* (there will be *no change* in our knowledge), because we already knew that the final state would always be $|1\rangle$ and never $|0\rangle$ – there is no alternative in this case. An equivalent thing happens if $p_0 = 1$. In either case there is no a priori uncertainty about the outcome – the novelty value of the result of a measurement would be zero. If on the other hand $p_1 < 1$ but still very close to 1, an occurrence of $|1\rangle$ would be greeted with a bored "So, what's new?" (very low novelty value), whereas if the rare state $|0\rangle$ were to happen, it would be an excited "Wow!" (very high novelty value). Finally, $p_0 = p_1 = 0.5$ should make us to expect the unity of average information gain, 1 b, and the highest novelty value, which we can arbitrarily set also equal to 1. For the general case of $p_1 \neq p_2$ the mathematical expressions for the novelty value of each particular outcome and the amount of information to be expected on the average before the operation of the pinball machine should reflect all these properties.

1.3 Quantifying Statistical Information

To find these expressions, let us call I_0 the information value or novelty value in bits if state $|0\rangle$ is seen, and I_1 the same for the occurrence of $|1\rangle$. We already stated that in the case of perfect symmetry, i.e., for $p_0 = p_1 = 0.5$, we should obtain $I_0 = I_1 = 1$ b. And it is reasonable to demand that for $p_i = 1$ ($i = 0$ or 1) we should get $I_i = 0$, whereas for $p_i \to 0$, $I_i \to \infty$. What happens in between? In general, the function we are looking for, $I(p)$, should be a continuous function, monotonically decreasing with p so that $I_i > I_k$ if $p_i < p_k$ (the value of the information gained should be greater for the less probable state). The following function fulfilling such conditions was chosen by *Shannon* and *Weaver* [104] for what is usually called the *information content* of an outcome that has the probability p_i to occur:

$$I_i = -K \ln p_i \, .$$

In order to obtain $I = 1\,\text{b}$ for $p_i = 0.5$ we must set $K = 1/\ln 2$. The negative sign is needed so that $I \geq 0$ (p always ≤ 1). Turning to logarithms of base 2, we can write

$$I_i = -\log_2 p_i \, .^\dagger \tag{1.1}$$

In Shannon and Weaver's original formulation the information in question would be a message that has been received from a predetermined set of possible messages – in our case of the pinball machine, it is the message that tells us that the final state was $|i\rangle$. It seems to me more appropriate and less confusing to always use the term *novelty value* for the function (1.1). For $p_i \to 0$, I_i tends to infinity (expressing the "Wow"-factor mentioned above when something very unlikely really does happen). Note that, in principle, the novelty value (1.1) has no relation to the subjective value of the outcome to the observer, which is difficult to quantify. In chance games, of course, such subjective value (e.g., the winning amount) is always set as a decreasing function of the p with the least probable outcome earning the highest amount. On the other hand, if the alternatives are possible outcomes of some natural event (e.g., weather, the behavior of a prey, etc.), the highest subjective value may well be that of the most probable outcome (the one expected based on subjective experience).

It is important to point out that the choice of the logarithmic form (1.1) does not emerge from any physical principle or statistical law. It is merely a *reasonable* choice that leads to useful relations (for instance, the choice $I = 1/p$ looks much simpler but it would be useless). Indeed, a logarithmic relation between I and p has fundamental mathematical advantages. Suppose that we consider two or more successive operations of our binary pinball machine and want to find out the total novelty value I_T of a specific set of successive outcomes of states $|a\rangle$, $|b\rangle$, ..., which have independent a priori probabilities p_a, p_b, ..., respectively (for instance, three consecutive $|0\rangle$; or the sequence $|0\rangle$, $|1\rangle$, $|0\rangle$...; etc.). In that case, we are asking for an overall occurrence whose probability according to elementary statistics is the *product* of the independent probabilities $p_a p_b \ldots$; therefore, according to (1.1): $I_T = -\log_2(p_a p_b \ldots) = I_a + I_b + \cdots$ In other words, the novelty value I defined in (1.1) is additive (any other functional form would not have this important property).

---

† Remember that, by definition, $\log_2 x = \ln x/\ln 2$.

The most important and useful quantity introduced in Shannon's theory is related to the question alluded to earlier: Given the probability values for each alternative, can we find an expression for the amount of information we expect to gain *on the average* before we actually determine the outcome? A completely equivalent question is: How much *prior uncertainty* do we have about the outcome? As suggested above, it is reasonable to choose the weighted average of I_0 and I_1 for the mathematical definition of the a priori average information gain or uncertainty measure H:

$$H = p_0 I_0 + p_1 I_1 = -p_0 \log_2 p_0 - p_1 \log_2 p_1 ,\ ^\dagger \qquad (1.2a)$$

in which $p_0 + p_1 = 1$. Note that H also represents something like the "expected average novelty value" of an outcome. Since H is a quantitative measure of the uncertainty of the state of a system, Shannon called it the *entropy* of the source of information (for reasons that will become apparent in Chap. 5). Note that the infinity of I_i when $p_i \to 0$ in expression (1.1) does not hurt: The corresponding term in (1.2a) tends to zero. Let us set, for our case $p = p_0$ (probability of finding the ball in the left bin); then $p_1 = 1 - p$ and:

$$H = -p \log_2 p - (1-p) \log_2 (1-p) . \qquad (1.2b)$$

Figure 1.2 shows a plot of H as a function of p (solid line). It reaches the maximum value of 1 b (maximum average information gain in one operation of our machine – or in one toss of a coin) if both probability values are the same ($p = 1/2$; symmetry of the machine, fair coin). If $p = 1$ or 0, we *already know* the result before we operate the machine, and the expected gain of information H will be zero – there is no a priori uncertainty. A measure of the average information available *before* we actually determine the result would be $1 - H$; $p = 1$ or 0 indeed gives 1 b of "prior knowledge," and $p = 1/2$ represents zero prior information (broken line), that is, maximum uncertainty.

We now expand our single-pin machine to a multiple one. For this purpose, we introduce additional pins in the way of each primary path, as shown in Fig. 1.3. In case of absolute symmetry, each possible initial trajectory will split into two equally probable ones, and the probability of finding the ball in one of the four lower bins will be 0.25. Since now there are four possible states, there is more initial uncertainty about the possible outcome and more information will be gained once we obtain a result. The final states can be labeled by the binary base-2 numbers 00, 01, 10 and 11, which in our familiar base-10 system are 0, 1, 2 and 3.‡

---

† Specific arguments supporting the choice of this logarithmic form are given in [102–104].

‡ Remember the general rule: If $D = \{x_{N-1} x_1 \ldots x_0\}$ is the binary notation of a number of N digits ($x_k = 0$ or 1), then $D = 2^{N-1} x_{N-1} + \cdots + 2^1 x_1 + 2^0 x_0 = \sum_{i=1}^{N} 2^{i-1} x_{i-1}$.

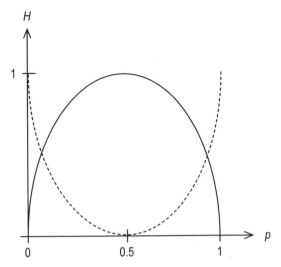

Fig. 1.2. Shannon's average information or entropy H as a function of the probability p of one of the final states of a binary (two-state) device, like the pinball machine of Fig. 1.1 (with some built-in bias). H is a measure of the uncertainty *before* any final state has occurred, and also expresses the average amount of information to be gained *after* the determination of the outcome. A maximum uncertainty of *one bit* (or maximum gain of information, once the result is known) exists when the two final states are equiprobable ($p = 0.5$). The *dotted curve* represents $1 - H$, an objective measure of the "prior knowledge" before operating the device

We can generalize the definition (1.1) for any number N of possible final states and do the same with relation (1.2), which will then read:

$$H = - \sum_{i=0}^{N-1} p_i \log_2 p_i \quad \text{with} \quad \sum p_i = 1 . \tag{1.3}$$

This function H has an absolute maximum when all p_i are equal, i.e., when there is no a priori bias about the possible outcome. In that case, by definition of the probability p_i, it is easy to verify that

$$H = \log_2 N . \tag{1.4}$$

This is a monotonically increasing function of the number N of equally probable states, indicating the increasing uncertainty about the outcome (actual determination of the final state) prior to the machine's operation, and the increasing amount of information expected to be harvested on the average once a measurement has been made. In the case of our expanded but still symmetrical machine with four equiprobable end states, we have $I_i = 2$ and $H = 2\,\text{b}$. A die is a system with six equally probable final states; the novelty value expected for each throw is, according to (1.1), $I = 2.58$; the expected average

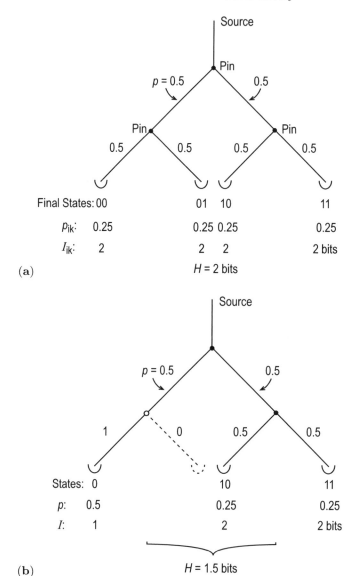

Fig. 1.3. **(a)** Multiple pinball machine with four equiprobable final states. To be identified, each state requires two binary digits (bits). The novelty value of any possible outcome when the machine is operated is also $I = 2\,\mathrm{b}$ and so is the expected information gain H. **(b)** Sketch of a pinball machine with final states of unequal probabilities 0.5, 0.0, 0.25 and 0.25, respectively. Occurrence of the first state has a novelty value of $1\,\mathrm{b}$ (lower, because it is more frequent), the others have two $2\,\mathrm{b}$ each (the second alternative does not count). The average information gain H is less than for an equiprobable, three-alternative case (which, for $p_i = 1/3$, would be $H = 1.59$)

gain of information or average initial uncertainty H is also 2.58 b. The higher the number of equiprobable alternatives, the greater the uncertainty before a measurement is made and the greater the value of the information once the actual result is known. On the other hand, if in relations (1.3) one of the final states has a probability $p = 1$, all other states are prohibited and there would be no information gained at all after a measurement. In general, for any p that is zero, the corresponding state is not an option: It does not count as an alternative (see second state in Fig. 1.3b). In general, if for a system with multiple alternatives some p turn zero for some external reason (i.e., if the number of end states with appreciable probabilities is small), the uncertainty will decrease, less information gain can be expected before a measurement is made, and less valuable will be a result on the average (but more valuable will be one of those low-p outcomes, relation (1.1)).

For a better understanding of the case of different probabilities, let us consider the pinball machine illustrated schematically in Fig. 1.3b with three actually possible final states. Let us assume that the pins are perfectly positioned so that their probabilities would be 0.5, 0.25 and 0.25, respectively. According to (1.1), occurrence of the first state represents a novelty value of 1 b of information, occurrence of either of the other two 2 b each (higher value, because they are half as frequent as the first one). The average information gain a single operation of the system can deliver is, according to (1.3), $H = 1.5$ b. It is less than that corresponding to Fig. 1.3a: The decrease is due to a decrease in the initial uncertainty, because in this case we do know that one of the four options will not occur.

Note that each node in the figure can be thought of as a binary machine – our expanded pinball machine is thus equivalent to a *network* of 1 b devices. If H_n is the expected average information gain for a device at node n, and P_n is the probability for that node to occur (= product of branch probabilities along the path to that node), it is easy to show, using relation (1.3), that

$$H_f = - \sum_{\text{final states}} p_f \log_2 p_f = \sum_{\text{nodes}} P_n H_n . \qquad (1.5)$$

p_f are the probabilities of the final states (= product of branch probabilities along the path to each final "bin").

A three-bit scheme is shown in Fig. 1.4a representing a system with eight possible final states; note the binary labels (code) assigned to each one. If the final states are equiprobable, i.e., if each one of the nodes represents a binary division with 50% probability for each branch, the novelty value of the result of a measurement will be, according to (1.1), $I_i = - \log_2 1/8 = 3$ b (= number of nodes to get to the bin in question). This is also the value of the entropy or average information gain H (1.4) in this case.

If on the other hand each node has a nonsymmetrical partition of probabilities as shown in Fig. 1.4b (displaced pins), the final states will have different a priori probabilities and the I_i values will be different like in Fig. 1.3b. To

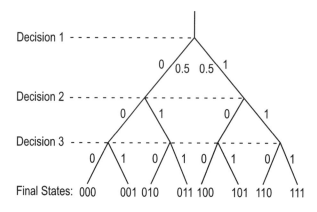

$I = - \log_2 (1/8) = 3$ bits each occurrence (novelty value)

(a) $H = 3$ bits

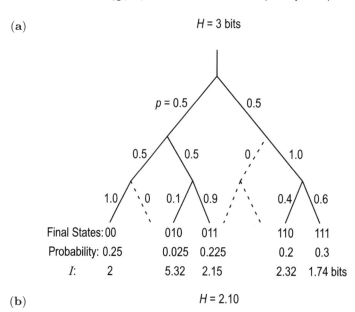

(b) $H = 2.10$

Fig. 1.4. (a) System with eight final states, all equiprobable. Notice carefully the labeling scheme, and how it relates to nodes and branches. In this case, $H = 3$. (b) Some "decision nodes" present two alternative paths with different probabilities, as indicated. Branches with zero probability must be taken out of the scheme: They represent no alternative or choice; final states with zero probability of occurrence do not count (whereas states with very small probability still do!)

calculate the a priori probability of a final state, we multiply the values of the branch probabilities along the path from S to the bin in question. As an example, in the figure we have assigned some specific probability values to each branch and show end-state probabilities and I values. What do we do with nodes like B and C that lead to a 0, 1 pair of probabilities (i.e., certainty of outcome at that node)? We just take them out of the scheme as nodes because they do not represent any alternative or choice (see also relation (1.5)). As mentioned above, end states with zero chance should be taken out, too. In our example, we are left with a system with only five final states. The corresponding value of H is 2.10 b. When all nodes have equiprobable branching (all $p = 0.5$), the probability of any final state is 2^{-N}, where N is the number of nodes to reach the final state.

There is another nonstatistical way of looking at diagrams like Fig. 1.4a. Consider each node as the fork in a road, i.e., a decision point with two alternatives. If your goal is to reach a given end state, you must use a map (have prior knowledge!) before branching off at each node. Suppose you want to reach the state 011 (this could be a location, a number, a letter, the item on a menu, a message, etc.), the map (the diagram in Fig. 1.4a, or a semantic description thereof) will allow you to make the right decisions and navigate until you have reached the goal. We can interpret the number of binary decisions (i.e., the number of nodes) as the amount of information needed to reach this goal, or as the amount of information "embodied" in the end state. For the equiprobable situation shown in Fig. 1.4a, the novelty value I_i (1.1) is indeed equal to the number N of nodes, i.e., the number of binary decisions that must be made to reach the i-th final state starting from the source ($I = -\log_2 2^{-N} = N$). Considering each end state generically as "a message," the expected average information H (1.4) is also called the *decision content* of the set of possible messages (a measure of how much choice is involved on the average in the selection of a message from that set). A diagram like shown in Fig. 1.4 is also called a *decision tree*.

Let us consider another example. Suppose I have a bowl with N little balls, 80% of which are black, and 20% white. I draw them out one by one and want to transmit the sequence obtained to somebody else. How many bits must I transmit *on the average,* if I were to repeat this process many times? In principle, one might expect that it would require one bit per ball, or N bits total. But whenever *all* balls of one color have come out, the color of the remaining balls will be known (and thus will not have to be transmitted), so on the average, less bits than N will be required. The quantity H takes care of that: Indeed, according to (1.2a) or Fig. 1.2, for $p_{\mathrm{black}} = 0.8$ (and $p_{\mathrm{white}} = 0.2$) we have $H = 0.72$, i.e., *on the average* only $0.72\,N$ bits will be required for a successful transmission of the string of N data.

Finally, as we shall see in Chap. 5, it is with relation (1.4) that the actual link with thermodynamic entropy is established: In a thermodynamic system in equilibrium, if W is the number of possible equiprobable microscopic states (number of possible equilibrium distributions of molecules in position and velocity compatible with the macroscopic variables), the Boltzmann entropy per molecule is defined as $s = k \log_2 W$.[†] Here comes a sometimes confusing issue. As we shall discuss in Sect. 5.3, in Boltzmann's statistical thermodynamics the entropy is a measure of our lack of knowledge, i.e., the uncertainty about a physical system's *microscopic* configuration: The more we know about its internal structure, the lower the entropy. A gas in a closed, insulated vessel whose molecules are all concentrated in one corner would have lower entropy than when they are uniformly spread over the entire volume, because in the first case we know far more (in relative terms) about the molecules' positions than in the second case. In fact, a uniform distribution of gas, as it happens when it is in thermodynamic equilibrium, will have maximum entropy (minimum "previous knowledge," like at the maximum of H in Fig. 1.2) compared to any other inhomogeneous distribution. In summary, an increase of knowledge about the microstructure of a system means a decrease in entropy and vice versa. But since more knowledge also means more information, Shannon's designation of H as entropy seems contradictory (this is why *Brillouin* [22] proposed the term "negentropy" – negative entropy – for H). However, note that in relation (1.2a) H refers to *potential* knowledge, i.e., it is an expression of the information an observer does not have but expects to gain on the average *after* he/she has learned of the outcome. And the expectation of receiving a maximum of new information after the measurement is equivalent to having maximum uncertainty (i.e., maximum entropy) *before* we make it. Notice carefully the time order involved: H measures the degree of uncertainty (entropy) of the system before its final state is identified in a measurement.

Let me rephrase all this: If we can predict the outcome of an event (that potentially has several alternatives) with certainty, we gain zero new information when the event happens – we already possessed this knowledge before. This means that there was no uncertainty and the average value of the outcome is, in effect, zero ($H = 0$). If we cannot predict the outcome, we have no prior knowledge – we can only gain it when the event happens and we observe the result; the *expected average amount* of new information is H. From a Boltzmann point of view, however, after the measurement has been made there will be a *collapse* of the entropy of the system because now we suddenly do have knowledge of the *exact* state of the system. Indeed, if the measurement were to be repeated a second time (assuming no other changes have occurred) there would be no prior uncertainty, and no new knowledge would be gained by carrying it out. Consider the subjective elements involved: First, the "prior knowledge" on part of the recipient (of the probabilities),

---

[†] H is dimensionless. A real equivalence with Boltzmann entropy requires that for the latter temperature be measured in units of energy (see Sect. 5.3), or that Shannon's entropy H (1.3) be redefined as $H = -k \sum p_i \ln p_i$ and expressed in J/K ($k = 1.380 \times 10^{-23}$ J/K is Boltzmann's constant, equal to R/N_{A}, the gas constant per mole divided by Avogadro's number). In the latter case, $1\,\mathrm{b} = k \ln 2\,\mathrm{J/K}$.

the decision of the sender (of setting the device), or the notion of "goal" or "purpose" when we refer to these decisions and the desire to inform the recipient. Second, if another observer comes in who does not know the previous outcome, the second measurement would deliver nonzero information to him/her – so we may state that the collapse of H really takes place in the brain of the beholder (a sudden change of the cognitive state of the brain). All this will play an important role in the pursuit of a truly *objective* definition of the concept of information in Chap. 3 – indeed, note that the concept of information *as such,* while used all the time, has not really been defined – only the amount of it!

1.4 Algorithmic Information, Complexity and Randomness

Consider now the binary number 1010111000101001101011100. This number is one out of $2^{25} = 33\,554\,432$ ways to order 25 binary digits. If we were to pick one string from that large set at random, the probability of getting an exact match would be 2^{-25} ($= 2.98 \times 10^{-8}$). Therefore, according to equation (1.4) the amount of information expected to be gained in one selection is, indeed, 25 b. There is another way of viewing this. The selection of each binary character represents one bit of information, because for each one we have made one choice to pick it out from two possibilities (0 or 1). So the total information content of the number is 25 b, because it consists of a spatial (or temporal) succession of 25 binary digits. Our "recipe" to represent, print or transmit that number requires 25 b or steps.

Intuitively, we may envisage the mathematical expression of the information content (not to confuse with the novelty value (1.1)) as something representing the "space" needed to store that information. In the case of a binary number, this is literally true. Now consider the sequence 1010101010101010101010101 (the number 22 369 621 in binary representation). It, too, has a 25 b information content and a probability of 2^{-25} to be picked out of a random collection of 25 binary digits. Yet we could think of a much shorter "recipe" for defining it, and a smaller space for storing it. The same happens even with an irrational number like π or $\sqrt{2}$, whose "recipe" can be expressed geometrically or by a numerical series. These examples tell us that sometimes information can be *compressed* – which means that instead of storing or transmitting each component character, we can store or transmit a short algorithm and regenerate the number every time we need it. This algorithm takes then the place of the whole sequence from the information content point of view. If the number of bits defining the algorithm is less than the number of bits defining the message, we can use it as a new quantitative expression of the amount of information. In most general terms, we can define *algorithmic information content* (of a number, an object, a message, etc.) as the shortest statement (measured in bits) that describes or generates the entity in question (for instance, [25, 116]). This applies to the labels of the final states of the case shown in Fig. 1.4a: We can give them quite diverse names for their identification, but listing three binary digits is the shortest way of doing it. The binary labels identifying an alternative are thus, algorithmic information. On the other hand, for a vibrating string with fixed end points, we can compress the information about its instantaneous shape by listing the (complex) Fourier coefficients (amplitudes and phases) of the vibration pattern up to a certain number given by the accuracy wanted in the description. A very complicated geometrical entity, like Mandelbrot fractals, may be generated with a very simple formula. The pattern or design generated by a cellular automaton is defined by a very simple rule or program, but would require a huge number of bits to be described or represented directly. And a simple spoken word can trigger a neural pattern in a brain which, or whose consequences, would require an incredibly large amount of information to be described (Sect. 4.2 and Chap. 6). In all this note that, in general, it is easier to generate a complex pattern from a simple "recipe," program or formula than the reverse operation, namely to find the algorithm that generates a given complex pattern – which in most cases may not be possible at all.

We stated in Sect. 1.1 that physics deals only with *models* of reality when it comes to quantifying a physical system and its dynamics (see Sect. 5.2). A model is an approximation, a mental construct in which a correspondence is established between a limited number of parameters that control the model and the values of certain physical magnitudes (degrees of freedom) pertinent to the system under study. So in effect we are *replacing* the physical system with something that is less than perfect but mathematically manageable – a set of algorithms. Is this the same as compressing information? Note a crucial difference: In the above example of the vibrating string, there is only one set of Fourier components that reproduces a particular shape of the string (within the prescribed accuracy) and no information is lost in the compression process; it is reversible and the original information can be recovered. The same applies to the relation between a Mandelbrot shape and its equation, or to a cellular automaton. In the case of the gas in thermodynamic equilibrium, however, the temperature as a function of the molecules' kinetic energies is an example where a huge amount of information is condensed into just one number – as happens whenever we take the average of a set of data (Sect. 1.1) – and there are zillions of different velocity arrangements that will lead to the *same* macroscopic temperature: The compression process is irreversible and a huge amount of microscopic information is indeed lost. In any kind of physical model, original information is lost in the approximations involved.

Quite generally, it is fair to say that physics (and natural science in general) deals mostly with algorithmic information. If among several physical magnitudes $R, S, U, V, W, \ldots$ there is a causal relationship between, say, variables R and S and the "independent" quantities $U, V, W, \ldots$, expressed by equations $R = R(U, V, W, \ldots)$ and $S = S(U, V, W, \ldots)$, the amount of information needed to specify a full set of values $R_k, S_k, U_k, V_k, W_k, \ldots$ is only that of the set $U_k, V_k, W_k, \ldots$ plus the amount needed to specify the *functional forms* for R and S. This would be the algorithmic information content of the full set, which could be much less than that needed for the full specification of the values of all variables. Another key example is that of the time evolution of physical magnitudes. In dynamics, one works with functional relationships of the type $R = R(U_0, V_0, W_0, \ldots, t)$, in which $U_0, V_0, W_0, \ldots$ are the values of pertinent magnitudes at some initial time t_0; again, we have the whole (in principle infinite) range of values of R before time t_0 and after t_0 defined by a limited set of initial conditions plus the bits necessary to define the functional relationship in question. In this sense, if the Universe were totally deterministic, each degree of freedom subjected to an equation of the type $R = R(t)$, its total algorithmic information content would remain constant, determined solely by its initial condition – the Big Bang. It is randomness which leads to gradual algorithmic information increase (see Chap. 3).

Algorithmic information also plays an important role in computer science. Like with information in Shannon's theory, rather than defining "algorithmic information per se" one can give a precise definition of the algorithmic information *content* without referring to any meaning and purpose of the information. Restricting the case to a string of binary digits one defines algorithmic information content as "the length of the shortest program that will cause a standard universal computer to print out the string and stop" (e.g., [42, 43]). This also applies to physics: The motion of a body subjected to given forces can be extremely complex – but if we know the equation of motion and the initial conditions, we can write a computer program that ultimately will print out the coordinates of the body for a given set of instants of time.

Algorithmic information can be linked to the concept of *complexity* of a system (such as a strand of digits, the genome, the branches of a tree, etc.): In principle, the more complex a system, the more algorithmic information is necessary to describe it quantitatively. Conversely, the more regular, homogeneous, symmetric and "predictable" a system, the smaller the amount of algorithmic information it carries. This is obvious from the definition of algorithmic information content given above. But there are examples for which a computer program (the "recipe") may be short but the computation process itself (the number of steps or cycles) takes a very long time. In such cases one may define a *time measure* of complexity as the time required for a standard universal computer to compute the string of numbers, print it out and stop. This is related to what is called *logical depth,* the number of steps necessary to come to a conclusion (compare with the concept of decision content in the preceding section), which is a measure of the complexity of a computational process. Algorithmic information can also serve as an expression of *randomness:* Looking at a sequence of binary characters, or at the objects pertaining to a limited set, we may call the sequence to be "random" if there exists *no* shorter algorithm to define the sequence than the enumeration of each one of its components.

The binary number sequences mentioned at the beginning of this section are only two from among a huge set – one is as probable (or improbable) to appear as the other when chosen at random from the entire set. It just happens that one appears more regular or organized than the other to us humans; here it is our brain that is applying some algorithm (pattern detection, Sect. 4.1 and Sect. 6.2). Yet when viewed as numbers expressed in base 10, both will appear to us as rather "ordinary" ones! Quite generally, the above "definitions" of complexity and randomness are not very satisfactory. First, how do we know that there is no shorter algorithm to define an apparently "random" sequence (as happens with the number 22 369 621 when it is changed to base 2)? Maybe we just have not found it yet! Second, both definitions would imply that complexity and randomness are linked: A gas in equilibrium (random from the microscopic point of view) would be immensely more complex than a Mandelbrot pattern (infinitely periodic, no randomness). Is this intuitively right? Third, it is preferable to treat complexity as a *relative* concept, as is done with Shannon information (expressing increase of knowledge), and deal with a measure of the *degree* of complexity rather than with complexity per se. This can be accomplished by turning to the existence of *regularities* and their probabilities of appearance, and defining "effective complexity" as the algorithmic information needed for the description of these regularities and their probabilities [8, 42, 43]. In this approach, the effective complexity of a totally random system like the molecules in a gas in equilibrium would be zero. On the other hand, it would also be zero for a completely regular entity, such as a string of zeros (a graph of effective complexity vs. randomness would qualitatively be quite similar to that shown in Fig. 1.2). We already mentioned that in a gas in equilibrium there are zillions of different states that are changing all the time but which all correspond to the same macroscopic state, described by average quantities such as temperature and pressure or state variables such as internal energy and entropy. Intuitively, we would like to see a requirement of *endurance,* permanence, stability or reproducibility associated with the concept of complexity, which does not exist for the microscopic description of a gas even if the macroscopic state is one of equilibrium. A crystal, on the other hand, does exhibit some degree of stability (as well as regularity), and so do strings of binary numbers and the atoms of a biological macromolecule. We can engender a crystal lattice with a formula (thus allowing for the quantitative determination of the algorithmic content), but we cannot do this for the instantaneous position and velocity of the molecules of a gas for practical reasons.

We mentioned the question of finding a shorter algorithm to describe a given string of digits and the relationship to randomness. Let us consider an example from biology. The DNA molecule contains the entire hereditary morphological and functional information necessary to generate an organism able to develop and survive during a certain time span in its environmental niche, and to multiply. The information is encoded in the order in which four chemical groups, the nucleotides, characterized by the four bases adenine (A), thymine (T), guanine (G) and cytosine (C), are linearly arranged in a mutually paired double strand (the famous "double helix"). This code-carrying sequence of nucleotides is called the genome of the species (there are a thousand nucleotides in the DNA of a virus, a few million in the DNA of a bacterium and more than a billion in that of humans). As we shall discuss in Sect. 4.4, the bases can be taken as the "letters" of an alphabet; since there are four possible ones (A, T, G, C) it takes two bits to identify each base of the genome (consider Fig. 1.3a). Thus, a sequence of n nucleotides requires $2n$ bits to be specified. But there are 4^n different ways to arrange such a sequence (a crazy number!), all *energetically equivalent* because there are no chemical bonds between neighboring nucleotides. Of these 4^n possibilities only a tiny subset is biologically meaningful, but there is *no* algorithm or rule known at the present time that would give us a "recipe" to identify that subset. In other words, we have no way of generating a genetic code with fewer bits or steps than just orderly listing each one of the millions-long sequence of bases found experimentally. Therefore, according to our first definition of random sequence, the genetic code does indeed seem "random." Yet, obviously, there is nothing random in it; we know that rules *must* exist because we know the biological *results* of DNA in action! Once we have understood how and why a particular sequence of nucleotides (really, a particular sequence of triplets thereof, see Sect. 4.4) is selected to code a specific protein, we might begin finding recipes for the biologically meaningful sequences and resort to algorithmic information as a shorter measure of the information content of the genetic code. Recently, some symmetries and regularities have indeed been identified [55]; we shall come back to this in more detail in Chap. 4. This is an example of the fact that, given a complex system with a certain Shannon information content, it is immensely more difficult to find out if it can be derived from some algorithm (i.e., determine its algorithmic information content), than to derive a complex system once an algorithm is given (as for instance in cellular automata).

One final point: We have been using numbers as examples. When is a number "information"? Base-2 numbers are most appropriate to express information content and to represent (encode) information (the alternatives, the messages). And we can define the algorithmic measure of a number in bits as the number of binary steps to generate it. But numbers are not *the* information – by looking at one, we cannot tell if it does represent information or not. In fact, a number only becomes information if it "does something," if it represents an action or interaction. This will indeed be our approach when we seek a more objective definition of information detached from any human intervention in Chap. 3. For instance consider a number L representing the length of an object. This number can express information only if the corresponding metadata (Sect. 1.1) are given: the unit of length used and the experimental error of the instrument. In particular, changing the unit will change the number to a different one: $L' = L\lambda$, where λ is the length of the old unit in terms of the new one. Yet L and L' express exactly the same thing (the length of a given object). Their algorithmic information *content* will be essentially the same because we know the rule of transormation from one to the other. Concerning the question of encoding information in a number, there is a famous example. If the distribution of decimal digits of π (or any other transcendental number) is truly random (suspected but not yet mathematically proven!), given any arbitrary finite sequence of whole numbers, that sequence would be included an infinite number of times in the decimal expansion of π. This means that if we were to encode Shakespeare's works, the Bible, this very book, or the entire written description of the Universerse in some numerical code, we would find the corresponding (long but still finite) string included in π! The problem, of course, would be to find out where it is hiding (there is no rule for that)! So, would this mean that π carries information – about everything? In a sense yes, but only because we *humans* have the capability of defining π operationally, defining a given sequence of integers (however long), and searching for it in the transcendental number!

1.5 The Classical Bit or "Cbit": A Prelude to Quantum Computing

In Sect. 1.3 our "standard binary pinball machine model" helped us introduce a measure for the novelty value of the outcome of an operation of the machine (relation (1.1)). We also defined a quantity H (1.3) which represents the information expected to be gained on the average by operating the machine and making a measurement to determine the final state. The unit of information is the bit, which we introduced as the quantity of information (new knowledge) expected after a single-pin machine with two equiprobable alternatives has been operated (or a fair coin has been tossed). We now ignore for a moment the operational side of our model and simply define as *a classical bit* the physical realization of a system – *any* system – that after appropriate preparation and operation 1. can exist in one of two clearly distinguishable, mutually exclusive, states (0–1; no–yes; true–false; on–off; left path–right path; etc.), and 2. can be *read-out* or *measured* to determine which of the two states it is (often called the "value" of the classical bit, which sometimes leads to a confusion of terms). If the two states are a priori equiprobable, the amount of information obtained in such a measurement will be one bit (see Fig. 1.2). This is the maximum amount of information obtainable from this binary device. Following *Mermin's* review article on quantum computing [76] we will adopt his nomenclature and call it "Cbit" to distinguish it from the quantum equivalent, the "Qbit" (commonly spelled *qubit* [100]), to be introduced in the next chapter.

Note that to evaluate the measurement result of a Cbit one has to have prior knowledge of the *probability* for a particular state to occur; only a statistical process would reveal the value of this probability. In computers and communication systems, a Cbit is usually set into one of the two possible states by an operation executed at the command of upstream elements to which it is connected (or by a human, if it is an input switch). It still may be viewed as a probabilistic entity, if the setting itself is not known to the observer or recipient at the time of measurement. This "premeasurement" vs. "postmeasurement" argument about "knowing" is important but tricky. We already encountered it in the discussion of the entropy (1.2) and it will appear again in full force when we discuss quantum information, as well as later in Chap. 3, when we turn to an objective definition of information. Let me just point out again that the concepts of "purpose" and "understanding" (of a message) are peeking through the clouds here! Although the Shannon information theory does not address such issues, information theory cannot escape them. At the roots is the notion of prior probability, a sort of metadata (Sect. 1.1) that must be *shared* between the sender of the information (the Cbit) and the receiver who carries out the corresponding measurement. Without such shared knowledge the receiver cannot determine the expected average amount of information, nor the novelty value of the information received.

We must emphasize that a Cbit is meant to be a *physical* device or *register,* not a unit of information or some other abstract mathematical entity. It is assumed to be a *stable* device: Once set, it will remain in the same state until it is reset and/or deliberately modified by some externally driven, specific, operation (e.g., the pin's position is reset, or the ball's paths are changed). Some fundamental operations on a single classical bit are the "identity" operation I (if the state is $|0\rangle$ or $|1\rangle$, leave it as is) and the "flip" operation X (if it is $|0\rangle$ or $|1\rangle$, turn it into $|1\rangle$ or $|0\rangle$, respectively – which can be accomplished physically, e.g., by inserting devices in Fig. 1.1 that make the paths cross). An example of an irreversible operation, which we will designate E_1, is "erase state $|1\rangle$ as an alternative" (which means that if $|0\rangle$, leave as is, if $|1\rangle$, set to 0), or similarly E_0, "erase state $|0\rangle$ as an alternative." Operation E_1 can be accomplished by an absorber in path 1, which means that the device will no longer provide any choice, i.e., it will always contain zero Shannon information. A mechanism that sets or resets a Cbit depending on some external input is called a *gate*. An apparently trivial, yet fundamental, fact is that a measurement to determine the state of a Cbit involves the identity operation ("leave as is"): It does not change the state of the Cbit. As anticipated earlier, this does *not* happen with a quantum bit (except when it is set in certain states – see next chapter).

So far we talked about the physical realization of a classical bit. It will be illuminating and particularly useful for the next chapter to also introduce a "mathematical realization," even if such formalism is unnecessary within the framework of the classical information theory. For that purpose one uses matrix algebra and represents a state by a unit vector (in an abstract two-dimensional space), such as:

$$|0\rangle = \begin{pmatrix} 1 \\ 0 \end{pmatrix} \quad |1\rangle = \begin{pmatrix} 0 \\ 1 \end{pmatrix} \tag{1.6}$$

Operations on Cbits are represented by matrix operators (e.g., see [76]). In general, the action of an operator $\boldsymbol{R}$ on a vector $|\psi\rangle$, described symbolically by the product $\boldsymbol{R}|\psi\rangle$, is a mathematical algorithm that changes this vector into another vector, $|\varphi\rangle$, in such a way that the result is independent of the frame of reference used to represent their components. These operators are called logic gates and one writes $\boldsymbol{R}|\psi\rangle = |\varphi\rangle$. For Cbits, the "leave as is" or identity operator $\boldsymbol{I}$, the "flip" operator $\boldsymbol{X}$ and an "erase" operator $\boldsymbol{E}_1$ would be represented in matrix form, respectively, as follows:

$$\boldsymbol{I}|0\rangle = \begin{pmatrix} 1 & 0 \\ 0 & 1 \end{pmatrix} \begin{pmatrix} 1 \\ 0 \end{pmatrix} = \begin{pmatrix} 1 \\ 0 \end{pmatrix} = |0\rangle$$

$$\boldsymbol{I}|1\rangle = \begin{pmatrix} 1 & 0 \\ 0 & 1 \end{pmatrix} \begin{pmatrix} 0 \\ 1 \end{pmatrix} = \begin{pmatrix} 0 \\ 1 \end{pmatrix} = |1\rangle$$

$$\boldsymbol{X}|0\rangle = \begin{pmatrix} 0 & 1 \\ 1 & 0 \end{pmatrix} \begin{pmatrix} 1 \\ 0 \end{pmatrix} = \begin{pmatrix} 0 \\ 1 \end{pmatrix} = |1\rangle$$

$$\boldsymbol{X}|1\rangle = \begin{pmatrix} 0 & 1 \\ 1 & 0 \end{pmatrix} \begin{pmatrix} 0 \\ 1 \end{pmatrix} = \begin{pmatrix} 1 \\ 0 \end{pmatrix} = |0\rangle \tag{1.7a}$$

$$\boldsymbol{E}_1|0\rangle = \begin{pmatrix} 1 & 0 \\ 0 & 0 \end{pmatrix} \begin{pmatrix} 1 \\ 0 \end{pmatrix} = \begin{pmatrix} 1 \\ 0 \end{pmatrix} = |0\rangle$$

$$\boldsymbol{E}_1|1\rangle = \begin{pmatrix} 1 & 0 \\ 0 & 0 \end{pmatrix} \begin{pmatrix} 0 \\ 1 \end{pmatrix} = 0 .$$

Examples of thus far physically meaningless operators, which, however, also will appear in our discussion of quantum information, are the following:

$$\boldsymbol{Y}|0\rangle = \begin{pmatrix} 0 & -1 \\ 1 & 0 \end{pmatrix} \begin{pmatrix} 1 \\ 0 \end{pmatrix} = \begin{pmatrix} 0 \\ 1 \end{pmatrix} = |1\rangle$$

$$\boldsymbol{Y}|1\rangle = \begin{pmatrix} 0 & -1 \\ 1 & 0 \end{pmatrix} \begin{pmatrix} 0 \\ 1 \end{pmatrix} = -\begin{pmatrix} 1 \\ 0 \end{pmatrix} = -|0\rangle$$

$$\boldsymbol{Z}|0\rangle = \begin{pmatrix} 1 & 0 \\ 0 & -1 \end{pmatrix} \begin{pmatrix} 1 \\ 0 \end{pmatrix} = \begin{pmatrix} 1 \\ 0 \end{pmatrix} = |0\rangle \tag{1.7b}$$

$$\boldsymbol{Z}|1\rangle = \begin{pmatrix} 1 & 0 \\ 0 & -1 \end{pmatrix} \begin{pmatrix} 0 \\ 1 \end{pmatrix} = -\begin{pmatrix} 0 \\ 1 \end{pmatrix} = -|1\rangle .$$

These operators are meaningless in the classical context because the states $-|0\rangle$ and $-|1\rangle$ do not represent anything physical for a classical Cbit.

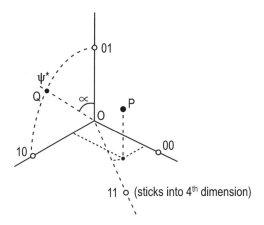

Fig. 1.5. Schematic view of the four vectors (1.8), each one of which represents a possible state of a classical 2-Cbit system (you are not supposed to see one of them, because it sticks out into the fourth dimension!). In classical information theory, the marked points are the only ones allowed in the 4D space tended by these unit vectors – any other point is meaningless. But in quantum information theory, other states (like points P and Q in the figure) are possible; they represent "entangled" states in which the great technological promise of quantum computing lies

We now extend our vector notation to two Cbits A and B. We again refer to Fig. 1.3, but view it not as the sketch of a physical model, but, rather, a flow chart. The two second-level nodes represent one and the same Cbit, associated with one or the other possible state of the first Cbit. What matters are the four final states in which a system of two classical bits can be found: $|0\rangle_A|0\rangle_B$, $|0\rangle_A|1\rangle_B$, $|1\rangle_A|0\rangle_B$ and $|1\rangle_A|1\rangle_B$. The state $|AB\rangle$ of the combined system AB can be described by the product of the two individual Cbit states:

$$\binom{1}{0}_A\binom{1}{0}_B = \begin{pmatrix}1\\0\\0\\0\end{pmatrix}_{AB} \quad \binom{1}{0}_A\binom{0}{1}_B = \begin{pmatrix}0\\1\\0\\0\end{pmatrix}_{AB}$$

$$\binom{0}{1}_A\binom{1}{0}_B = \begin{pmatrix}0\\0\\1\\0\end{pmatrix}_{AB} \quad \binom{0}{1}_A\binom{0}{1}_B = \begin{pmatrix}0\\0\\0\\1\end{pmatrix}_{AB} .$$

(1.8)

These are the only possible states in which a 2-bit classical system can be found; we have represented them schematically in Fig. 1.5. Any measurement will yield, obviously, two bits worth of information. The measurement of a register consisting of n Cbits will provide only *one* result out of n possible ones and the average information value (1.4) (entropy) of a measurement is n bits if all bifurcations are equiprobable (otherwise it would be less). Moreover, it is important to remark that any measurement will also provide information on each component Cbit (Fig. 1.3), which has a well-defined state of its own. Indeed, in the classical case, component Cbits preserve their individuality; a measurement does not change their states (nor that of the combined 2-bit or n-bit system) and there is no "hanky-panky" between different Cbits; changing one will not change the partners – there is *no* active entanglement! All this does not apply in the quantum case!

1.6 Objective and Subjective Aspects of Classical Information Theory

How does all the above relate to our principal aim, namely to find an objective definition of information that embodies its major properties and roles in Nature in a way independent of human, subjective, biases regarding its use? Consider a complicated Chinese character: Shannon information or algorithmic information deal quantitatively with the sign itself, i.e., with its representation as a pattern (say, in the arrangement of black-and-white pixels, or in some rule on how to draw it). But they do not deal with why it is there, what it is intended to do and what it means to anyone who sees it, and understands it or not. That is the kind of information we are after – one that brings out the *relation* between intention and effect! Furthermore, we have asked in the Introduction: Is there a relationship between information and complexity? Evidently not in terms of Shannon or algorithmic information when it comes to intention or effect: Something very low in bit content as the word "yes" could have extraordinarily complex consequences, such as a marriage or the detonation of a nuclear bomb!

The standard pinball machine and the Cbits referred to in this section to illustrate some of the fundamental tenets of Shannon's information theory are human-designed artifacts with a purpose of human interest. This applies to the classical bit and all elements considered and used in traditional information theory. But what about the Universe as a whole in absence of humans? Does classical information theory apply to what happened before we were around to ask questions about the outcome of possible alternatives? And what about biological systems: Can Shannon's theory adequately describe what happens with living systems when it comes to information? We already stated in the previous section that each one of the 4^n different, energetically equivalent, ways to arrange the sequence of bases in a given DNA molecule (an impossibly big number) has the *same* amount of Shannon information! Yet only an incredibly tiny subset is biologically significant. It should be clear from this example that the amount of information as defined in relation (1.4) only refers to *structural* information, such as the number of steps needed to assemble the nucleotide chain of a given genome, be it chemically in a "primeval soup" or symbolically for storage in a computer. It has nothing to do with the actual biological purpose or function of the DNA, namely to serve as a blueprint for the formation and control of a living organism capable of surviving and multiplying in its ecological niche.

The concept of information we have been using and expressing quantitatively in Sect. 1.3 and Sect. 1.4 (without formally defining it) does have some of the characteristics we are seeking for a more general, dynamic concept of information (see Introduction). For instance, whenever we stated above that "we learn," "we have knowledge of," "we look into," "we measure," we were, indeed, referring to a *change* triggered in our brain by some external pattern perceived by our senses: a change of the state of our own knowledge (ball in left bin or in right bin, heads or tails after the toss, "state 010," etc.). But what a complicated system and processes are we referring to! As a matter of fact, we will have to wait until the last chapter of this book, to even only *speculate* about the complex physical-physiological processes that code sensory information and "represent" knowledge in the brain! Even if the change is one occurring in some inanimate registering apparatus, we must not forget that the latter was devised by a human being for the very purpose of eventually changing somebody's state of knowledge. Again, traditional information theory only deals with relations and quantities describing amount of information, leaving aside all questions about the meaning and ultimate purpose of the information that is being dealt with. As a consequence, this theory is restricted to a limited domain of applications – albeit crucially important ones because of their relevance to technology.

Another question revolves around the fact that so far we have been dealing with systems (our idealized pinball machine, or, in general, the Cbit) that only assume *discrete* states. They are human artifacts, created for a specific purpose, and the question arises: Are there also natural nonliving systems that can be found only in discrete states? Most macroscopic variables that we use to describe the inanimate objects that surround us vary continuously, with relatively few exceptions (for instance, material phase changes are discrete transitions of the state of a body), and it is quite difficult to find a natural macroscopic nonliving equivalent of our pinball machine or a network of Cbits (see, however, our discussion of the measurement of a classical macroscopic variable as a process that proceeds in binary steps, Sect. 5.5). Discrete states do appear, however, when humans and animals are confronted with decisions about environmental phenomena that present them with alternatives requiring binary yes/no decisions. In the microscopic world, where systems are *quantized,* there are discrete states for many physical variables. We will, indeed, find natural equivalents of our pinball machine in the quantum world (although some human intervention will be necessary there, too – see next chapter); however, their behavior turns out to be rather "bizarre"!

Let me conclude this chapter by explicitly confronting the objective aspects of classical information theory with the subjective aspects that appeared on several occasions in this chapter. In Table 1.1 we attempt to summarize the basic physical, mathematical and subjective points. Classical information theory, of course, only deals with the first two columns. The items in the third and fourth columns will be introduced gradually in the following chapters, and will be discussed in detail in Chap. 6.

Table 1.1. Basic objective and subjective processes intervening in classical information theory. The relations between items in columns 1 and 2 were described in the present chapter. Relations between columns 3 and 4 will be discussed in Chap. 6 and, partly, Chap. 4. The ultimate *purpose* of the actions listed in column 1 is to elicit the corresponding reactions listed in column 4

Physical systems	Mathematical expressions	Subjective links	Neural correlates
Properties of system and possible alternatives		Motivation to select or define system	Cortico–limbic interaction: setting goal and planning use
Set initial state and set (or determine) prior probabilities of outcome	*Novelty value* of a given alternative: $I_i = -\log_2 p_i$; *Entropy* (degree of unpredictability or expected average information gain): $H_{\text{before}} = -\sum p_i \log_2 p_i$	Initial knowledge about the system: expected value and uncertainty	Cognitive (cortical) state of brain before activation/observation of the system: expectation, imagination of result
Operate the system		Motivation to acquire knowledge about the outcome	Cortico–limbic interaction driving actions needed for operation and/or observation
Measurement/ determine outcome	Statement about outcome, *Collapse of entropy* (no uncertainty): $H_{\text{after}} = 0$	Gain of new knowledge	Change of cognitive (cortical) state of brain

$\underbrace{\qquad\qquad\qquad\qquad\qquad\qquad}$
Shanon information theory

$\underbrace{\qquad\qquad\qquad\qquad\qquad\qquad\qquad\qquad}$
Pragmatic information link

2 Elements of Quantum Information Theory

The previous chapter dealt with classical information, Shannon's theory and Cbits, a generic name for devices which have two possible, mutually exclusive, states, each one of which has a certain a priori probability to occur. Normally, we think of a Cbit as a "symmetric" device, in which both states are equiprobable. A Cbit has to be prepared and operated (in our model of a binary pinball machine, the ball released, or in a coin toss, the coin thrown), or set by a gate (the usual mode of operation in a computer, which fixes the state in which the Cbit will be found as a function of some input signal), and it has to be "measured" or read out (which will reveal the actual state in which it has been left). Cbit devices are fundamental components of information processing, communication and control systems. In modern technology they function on the basis of quantum processes (semiconductivity, photoelectric effect, tunneling effect, etc.) but they are not in themselves quantum systems: Rather than with single photons or individual electrons, they operate with classical streams of particles (e.g., pulses of coherent electromagnetic waves in optical networks and electric currents in conductors and semiconductors). In recent years, however, the possibility of using quantum devices has opened up new possibilities for a vastly different type of information processing, although some physicists have reservations about a speedy industrial-scale implementation.

Classical information theory has been around for over 60 years and there are hundreds of well-tested textbooks not only for physics and math students but also for biologists, engineers and economists. In contrast, quantum information theory is in its infancy – I should say in its gestation period – and it involves physics concepts that are not necessarily familiar to everybody. So before we discuss some fundamental concepts of quantum information we should review some of the basics of quantum physics for the benefit of readers less familiar with the subject. True, for someone only interested in the practical aspects of quantum computing, intimate knowledge of quantum mechanics is not necessary, just as knowledge of condensed matter physics is not required for a computer programmer [76]. However, our little primer on quantum physics will also be useful for later discussions of the concept of information, especially in Chap. 5.

2.1 Basic Facts about Quantum Mechanics

What is a quantum system? An easy way out is to say that it is a microscopic system in which the laws of classical physics derived from the study of macroscopic bodies break down. Quantum systems obey the laws of *quantum mechanics* (QM). Since macroscopic bodies are made up of huge ensembles of quantum systems, the equations of classical physics must in some way be derivable from QM in the limit of certain domains and/or number of particles exceeding some critical scale-size – this is called the correspondence principle or Ehrenfest's theorem. We should state at the outset, however, that there is no sharp, unambiguously defined limit between the two domains – the issue of quantum–classical transition still remains one of the "twilight zones" of physics, presently under vigorous investigation and discussion (theory of decoherence, e.g., see [117] and Sect. 5.6).

The most fundamental difference between a classical and a quantum system is that the latter cannot be observed (measured) without being perturbed in a fundamental way. Expressed in more precise terms, there is no process that can reveal any information about the state of a quantum system without disturbing it irrevocably. Of course, classical systems can also be disturbed by a measurement (think of what would happen if you were to measure the temperature of a small volume of warm liquid with a large cold mercury thermometer!). But for the macroscopic world there is the fundamental assumption that such disturbance is merely "accidental," due to the imperfection or inadequacy of the instruments used, and that it is possible, at least in principle, to gradually improve the measurement method to make any perturbation as small as desired, or, at least, develop corrective algorithms to take these perturbations into account. Quantum systems cannot be left undisturbed by measurement, no matter how ideal the instruments are: There are intrinsic limitations to the accuracy with which the values of certain physical magnitudes or "observables," as they are called in QM, can be determined in measurements. Moreover, the type of measurement itself (i.e., the corresponding experimental setup) dictates the integral behavior of the system even before the actual act of measurement has taken place (unless the system is in an eigenstate of that observable; see below) – for instance, it determines whether an electron or a photon will exhibit a particle-like or wave-like behavior.

The intrinsic limitation to our potential knowledge of a quantum system is most concisely expressed in the form of the *Heisenberg uncertainty principle*. For a single particle traveling along the x-axis with momentum p_x, this principle states that

$$\Delta x \Delta p_x \geq \frac{\hbar}{2} \, . \tag{2.1a}$$

Δx and Δp_x are the standard deviations of measured values of position and momentum, respectively, obtained for a given type of particle in a series of experiments under strictly identical circumstances of preparation (experimental setup and initial conditions) and measurement (instrumentation and timing). According to the meaning of standard deviation, Δx and Δp_x represent the approximate ranges within which the values of the position and momentum can be expected to be found with "reasonable" probability (68% for a Gaussian distribution) if measured under the specified conditions. The constant $\hbar = h/2\pi = 1.05 \times 10^{-34}\,\mathrm{m}^2\cdot\mathrm{kg}\cdot\mathrm{s}^{-1}$, where h is the famous *Planck constant*. Similar expressions exist for the standard deviations of all coordinates and corresponding components of the momentum vector. We should emphasize that the restriction (2.1a) only applies to coordinate and momentum components along the *same* axis; no such limitation exists between, say, Δx and Δp_z.

Relation (2.1a) tells us that the more accurate our knowledge of the position x of a particle (the smaller Δx), the larger the uncertainty (Δp_x) of its momentum, and vice versa. For the macroscopic world, this is of little consequence: Plugging in a reasonable value for the error Δx of position measurements of a macroscopic body would yield numbers for Δp_x absolutely negligible compared to the normal errors of momentum (or velocity and mass) measurements. So, clearly, h is a universal constant that has to do with the scale-size of the quantum-classical transition mentioned above.

In the quantum domain the consequences of (2.1a) are far-reaching. Indeed, the key fact is that this uncertainty does not only refer to *our* knowledge or "knowability" of the coordinates and the momentum vector of a particle, but that it has *physical consequences* for the behavior of the particle itself. In other words, an experimental setup to measure one of the two variables (say, x) with high accuracy, or to physically restrict its domain of variability (which is equivalent, because it imposes an a priori upper limit to the spatial uncertainty), will have physical consequences for the dynamic behavior of the particle even if we do not look at the measurement results. For instance, if we place a barrier with an open slit of aperture Δy perpendicular to a particle's motion along x, according to (2.1a) the corresponding component of the momentum p_y will exhibit an average spread of $\Delta p_y \geq \hbar/(2\Delta y)$: The particle's motion will suffer an unpredictable kink, i.e., be diffracted at the slit, and after many repetitions of the experiment under exactly the same conditions, we will have generated an ensemble of impacts arranged in a diffraction pattern on a detecting screen: The particle shows the behavior of a wave! Notice that this seems to contradict the very existence of the concept of a "trajectory" of a particle. However, this concept is classical in the sense that to call something a trajectory, we must determine it "point by point" in an appropriate macroscopic experimental setup such as a photographic emulsion, the fluid in a bubble chamber, multi-layers of counters, etc. In such devices the experimental uncertainty Δy perpendicular to the direction of motion, a classical domain error, is very big compared to a quantum domain's linear dimension, and the associated Δp_y will be extremely small compared to the momentum p_x in the direction of x. This means that the direction of motion will not be altered visibly (for a charged particle the deviation will be infinitesimal compared to its average Coulomb scattering in the medium). In an undisturbed quantum system left to itself there are no defined point-by-point trajectories (see also Sect. 2.3 and Sect. 2.5).

A relationship equivalent to (2.1a) links the uncertainties in energy E and time t:

$$\Delta E \Delta t \geq \frac{\hbar}{2} . \tag{2.1b}$$

This expression, too, has important consequences for quantum systems. For instance, it tells us that the principle of conservation of energy can be violated by an amount ΔE during short intervals of duration $\approx \Delta t$ determined through relation (2.1b). For classical, macroscopic systems the interval Δt would be totally negligible even for the smallest classical ΔE. But in the quantum domain, a photon that does not have the necessary energy can "borrow" some extra energy and create a "virtual" electron–positron pair (minimum cost: $\Delta E = 2 \times m_e c^2 = 1022 \, \mathrm{keV}^{\dagger}$) that after an extraordinarily short interval of time given by (2.1b) (so that it cannot be observed directly – that is why it is called virtual) annihilates itself to restitute the photon (see Sect. 3.4).

Another important equation (de Broglie's relation) expresses the earlier mentioned duality between particle (of momentum p) and wave (of wavelength λ):

$$\lambda = \frac{h}{p} . \tag{2.2}$$

Any particle of momentum p has a wave associated, of wavelength λ. As stated above, it really depends on the type of measurement whether it will be seen to behave like a particle or, instead, a wave. Now, when we say "a particle of momentum p" we are implying that $\Delta p \approx 0$, i.e., for a one-dimensional case, according to (2.1a), $\Delta x \to \infty$, which means that we cannot determine the location of the particle at all: It would be a "pure wave" extending from minus to plus infinity. In a real case, however, we will have a "wave packet" of spatial extension Δx (outside of which the amplitude of the wave $\to 0$). According to Fourier analysis this implies a superposition of waves with a continuous spectrum of frequencies or wavelengths within a certain interval $\Delta \lambda$, which according to (2.2) also means a range of momentum values of a finite extension Δp. It can be shown mathematically that in this case both Δx and Δp_x satisfy relation (2.1a); this relation, therefore, appears as a mathematical consequence of the dual nature of particles expressed by (2.2).

The smallness of Planck's constant gives an idea of the smallness of the quantum domain (the wavelength of a typical TV cathode ray tube electron

---

† $1 \, \mathrm{keV} = 1000 \, \mathrm{eV} = 1.602 \times 10^{-16} \, \mathrm{J}$.

is $\sim 10^{-9}$ m). But in general it is quite difficult to place an actual number to quantum dimensions – it will depend on the system to a certain degree. Large organic molecules like proteins (tens of nanometer size) behave classically in many respects and can be manipulated as such, although their internal structure and their interactions with other molecules are governed by QM. Diffraction experiments have been carried out successfully with C_{60} fullerene molecules ("buckyballs") showing their wave-like quantum behavior under such experimental conditions. The oscillations of heavy macroscopic low-temperature test bodies for the detection of gravitational waves must be treated as a quantum phenomenon (e.g., [18]). On the other hand, there are collective quantum effects that transcend into the classical, macroscopic domain, such as the behavior of certain fluids near absolute zero temperature, or ferromagnetism. And we already have stated that any present-day computer works on the basis of quantum processes, yet it is an eminently classical device. A conceptually more appropriate way is to talk about domains of quantum *interactions:* Experimental setups are *all* classical at one point or another, of necessity (or our classical sensory system would not be able to extract any information – see detailed discussion of the measurement process later, especially in Chap. 5), but they contain domains in which the quantum systems under study interact with each other and with quantum components of the experimental setup.

We have talked about the wave-like behavior of particles; now we shall consider the "particle-like" behavior of electromagnetic waves and write down three important relations valid for photons (this is really how quantum physics began historically!). The energy E and frequency ν of a photon (zero rest mass traveling in vacuum with the invariant speed $c = 2.998 \times 10^8$ m/s) is given by Planck's formula:

$$E = h\nu = \hbar\omega \,. \tag{2.3a}$$

($\omega = 2\pi\nu$ is the angular frequency.) If j is the flux of photons in a coherent light beam (number of photons per unit surface and time), and $S = EB/\mu_0$ the magnitude of the Poynting vector (electromagnetic energy flux per unit surface and time), according to (2.3a) we have

$$j = \frac{S}{\hbar\omega} \,. \tag{2.3b}$$

This is a relation that links classical electromagnetism (the continuous quantity S related to the amplitude squared of a wave) with the quantization of electromagnetic radiation (the discontinuous quantity j related to the number of particles per unit surface and time). Given a source of radiation (light source, antenna), the transition occurs when the radiated power is turned down to about the energy of one photon ($\hbar\omega$) per minimum detectable interval of time. In that limit, j in relation (2.3b) must be understood as being a temporal *average value*. The momentum of a photon can be calculated from de Broglie's relation (2.2):

$$p = \frac{h\nu}{c} = \frac{h}{\lambda}. \qquad (2.4)$$

In anticipation of things to come, let us state here that the entire frame of QM can be developed from just a few basic principles (e.g., [11, 32, 65]) of which relations (2.1) represent a special case. The other relations thus far discussed are consequences of these principles.

The formalism of QM is quite different from that of classical mechanics, working with mathematical representations of two fundamental entities: 1. the state of a system, symbolized by a vector $|\psi\rangle$ in an abstract Hilbert space, and 2. a set of operators, symbolized here by S, representing both the observables involved and the transformations imposed on the state of a system by interactions with other physical systems. When continuous variables such as coordinates are involved, the state is represented by a continuous function – the quantum *wave function*.[†] On the other hand, there are situations in which we can ignore most degrees of freedom of a system (the variables that define its full state at any given time) and limit its description to a small number of them. This happens when our attention is focused on some discrete "internal" states such the spin of an electron (up or down along a given axis), the state of polarization of a photon (vertical or horizontal), or "external" average spatial states such as the particular route taken by a particle from among two possible ones (e.g., "left path" or "right path"). In such a case we can represent the state of a system by a low-dimensional vector – the quantum *state vector* (for instance, as expressed by relation (1.6)).

For the limited scope of this book, we shall describe only briefly the general mathematical framework on which QM is built. Given a quantum system (which means given some interacting physical entities such as particles and fields and pertinent initial and boundary conditions), and given an observable S (which we consider to be defined by its measurement procedure), one finds that there exist certain states of the system for which the measurement of that observable consistently yields the *same* value (a real number). These states, called *eigenstates* or *eigenfunctions* of the observable, in many cases form a discrete enumerable set which we shall designate with $|\varphi_k\rangle$; the corresponding values s_k of the observable are called its *eigenvalues;* since they represent measurement results, they must be real numbers.

One fundamental task of QM is to set up algorithms that allow the calculation of such eigenstates and eigenvalues. This is done by casting the

---

[†] This is not to be confused with an ordinary traveling wave, whose amplitude describes intensity or energy density, and which carries energy and momentum from one point to another at a finite speed – none of these properties apply to a quantum wave function. The wave function *is* – it is an abstract nonlocal representation of the state of a system which does not propagate from one place to another, although from it one can obtain information on the time-dependent probability distribution of moving particles (represented by propagating wave packets). Causality applies only to macroscopically measurable effects (see Sect. 2.9 and Sect. 5.5).

operators and state vectors or wave functions in such a mathematical form that the wanted entities are solutions of the following type of equation:

$$S|\varphi_k\rangle = s_k|\varphi_k\rangle \, .$$

As noted above, when the observable corresponds to a continuous variable, S will in general be a continuous function or a differential operator and the discrete set of $|\varphi_k\rangle$ will be functions; if S represents a discrete quantum observable (see Sect. 2.4), it will be a finite matrix and the eigenfunctions will be a limited number of vectors. The actual form of the operator S is determined by physical considerations concerning the measurement process which defines the observable in question. Two observables like x and p_z (called commuting observables) that are simultaneously measurable without being subjected to an uncertainty relation, have common eigenfunctions. All operators that represent physical quantities must have real eigenvalues (and have some other specific properties some of which will be considered later). In Sect. 2.4 we will discuss a few concrete applications of this latter case. Note that, almost by definition, when the state of a quantum system is the eigenstate of an observable, the system will behave classically with respect to *that* observable: A measurement will not induce any change – we will always obtain the same value of the observable and the state of the system will remain unaffected by the measurement. We call such state a *basis state;* when a system is in a basis state, eigenstate of an observable, a measurement of that observable will give a definite result with certainty (the corresponding eigenvalue). In that case a measurement provides zero Shannon information (see Sect. 1.3).

Eigenstates (basis states that form what in mathematics is called a complete orthogonal set) represent a very special subset of all possible states of a quantum system. One of the fundamental principles of QM states that the most general state of a system $|\psi\rangle$ (prepared in a known, but otherwise arbitrary, way) can always be expressed as a *linear* superposition of eigenstates (principle of superposition, or "Max Born rule"):

$$|\psi\rangle = \sum_k c_k|\varphi_k\rangle^\dagger \tag{2.5a}$$

in which the amplitudes c_k are complex numbers (we will explain later why they must be complex numbers). When the system is in such a *superposed* state, the measurement result of the observable S still will be one of its eigenvalues – no other value is allowed under any circumstances (the quintessence of "quantification"!). It could be any one of the set; what the particular state $|\psi\rangle$ defines is the *probability* p_k with which a particular eigenvalue s_k will appear (in identical measurements of identical systems prepared in identical ways). Moreover, once a measurement is made, the system will emerge

---

† The $|\varphi_k\rangle$ are *unit* vectors (basis states are always considered unit vectors); the c_k are called the *projections* of the state vector $|\psi\rangle$ onto the basis vectors; $c_k|\varphi_k\rangle$ are *components* of the state $|\psi\rangle$.

from the process in the eigenstate $|\varphi_k\rangle$ that corresponds to the eigenvalue obtained – in other words, a measurement process *changes* the initial superposition (2.5a) and leaves the system in an eigenstate. This is called a "collapse" of the wave function or state vector; it is an irreversible process: Once it happens, all information on the initial state (the c_ks) is wiped out and the original state cannot be reconstructed.[†] Traditional quantum mechanics does not address what "exactly" happens *during* the collapse; this issue is still very much under debate. Which eigenvalue s_k and corresponding eigenstate will emerge as the result of a measurement cannot be predicted – the probability with which it will be obtained is given by the real number

$$p_k = c_k c_k^* = |c_k|^2 \tag{2.5b}$$

where c_k^* indicates complex conjugate. This requires a *normalization condition* for the amplitudes c_k:

$$\sum_k c_k c_k^* = \sum_k p_k = 1 . \tag{2.5c}$$

Only an *exact* repetition of the experiment many times will provide statistical information on the probabilities p_k. Note that if we multiply the state vector or state function $|\psi\rangle$ by a constant, we will obtain a quantity that represents the same state, i.e., will have the same physically verifiable properties. However, when a state vector is part of a superposition like (2.5a), multiplication by a real or complex number will alter the total state. When the system is in a superposed state (2.5a), the expectation value of an observable S will be given by $\langle S \rangle = \sum_k p_k s_k = \sum_k |c_k|^2 s_k$. It is clear that for a given observable the probabilities p_k will change as the state of the system (2.5a) changes. Concerning relation (2.5a), if the system is in one of the eigenfunctions of an observable, $|\psi\rangle \equiv |\varphi_i\rangle$, all c_k will be zero and $c_i \equiv 1$, $p_i = 1$. But note carefully that only the observable of which $|\varphi_i\rangle$ is an eigenstate will give a known result with certainty (zero new information); this in general will not be true for other observables. It is instructive to apply here concepts discussed in Sect. 1.3. The a priori average amount of Shannon information to be obtained in a measurement will be, according to relation (1.3), $H = -\sum_k p_k \log_2 p_k$. Note that if the system is initially in an eigenstate, one of the probabilities will be equal 1 and $H = 0$. The maximum obtainable information about an observable is $H_{\max} = \log_2 N$ (relation (1.4)), where N is the number of corresponding eigenstates (which could well be infinite).

There are interaction processes that change a quantum system from one state $|\psi\rangle$ to another $|\psi\rangle'$. These are transformations that can be continuous and reversible (caused by interactions with other quantum systems) or discontinuous and irreversible (caused by interactions with macroscopic, classical devices); they are represented by operators, too. We need mathematical

---

[†] There are different kinds of collapses, including reversible ones (Sect. 5.5).

relationships between transformation operators and those corresponding to observables in order to represent the physical processes that control the evolution of a quantum system. These relationships are drawn from classical physical laws which are "translated" into equivalent relationships between the corresponding operators – this is an expression of the correspondence principle mentioned at the beginning of this section.[†] One such relationship (derived from the Hamilton–Jacobi formulation of classical mechanics) leads to the famous Schrödinger equation (Sect. 5.6) which describes how the state of a quantum system evolves as a function of time. Quite generally, we should mention that in this evolution a system behaves in a strictly deterministic and continuous way as long as it remains a closed system in its own quantum domain (see Sect. 5.3). The moment an interaction or chain of interactions with the macroscopic (classical) world takes place, be this the purpose-directed, organized action of a measurement or some stochastic "fuzzy" interaction with the outer environment (Sect. 5.5), any interference (superposition) will be destroyed (that is the reason why this is called decoherence) [117].

The following four interconnected features are unique to the quantum world:

1. the fact that quantum systems are deterministic only when left to themselves, unobserved, free from other interactions with macroscopic systems;
2. the existence of superposed states, which can be interpreted as a system being in different states "at the same time";
3. the restricted repertoire of alternatives for the values that can be obtained in the measurement of observables; and
4. the irreversible collapse of the general state of a system when a measurement is made (i.e., when the "system" is no longer the original one but the expanded "system under measurement" – see Chap. 5).

These features are quite germane to quantum computing; we shall encounter them throughout the following sections and then again in Chap. 5.

2.2 Playing with a Quantum Pinball Machine

We now introduce the "quantum binary pinball machine" shown in Fig. 2.1. The "balls" are photons emitted by a source S of coherent light (e.g., a monochromatic laser) that can emit individual particles of wavelengths in a narrow interval. The "pin" is a half-silvered semireflecting lossless mirror B_1, generically called a *beam splitter,* and the "bins" are photon detectors D_0 and D_1. Note that our quantum pinball machine is a classical artifact with some

---

[†] There are quantum observables that do not have classical equivalents, like the spin. For these, other kinds of relationships are set up, which we will not consider here.

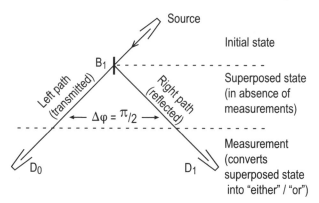

Fig. 2.1. Quantum "binary pinball machine." A source S of coherent monochromatic light sends a beam of photons to a symmetric nonabsorbing semireflecting mirror B_1. Half of the beam intensity is transmitted, the other half is reflected and phase-shifted by $\frac{\pi}{2}$ with respect to the transmitted one. D_0 and D_1 are photon detectors

well-defined quantum domains in which some specific transformations (half-silvered mirrors) and interactions (the detectors) are performed on quantum particles (photons or other particles).

First we study the device's properties using a "classical" continuous beam of coherent light emitted by the source S. A symmetric semireflecting non-absorbing mirror splits the light beam into two: One part is transmitted through at half the original intensity, the other part is reflected, also at half intensity. Furthermore, it can be demonstrated [30] that in this case the phase of the reflected beam (on the right side in our example) is shifted by $\varphi = +\frac{\pi}{2}$ (a quarter wavelength) with respect to that of the transmitted beam.

Now we decrease the source intensity steadily. If there was no quantum physics, we could do so until the intensity becomes zero (the electromagnetic field representing the wave vanishes). But according to (2.3a) and (2.3b) this model of a continuous stream of electromagnetic energy breaks down, almost like a steady stream of water from a faucet will break into droplets as we gradually decrease the flow. Eventually we will reach the situation of sending one single photon at a time (this can be accomplished now readily in the laboratory). What will happen?

One finds that for each photon injected, either D_0 or D_1 will respond, with a 50% probability each. The distribution of responses will be random – a result identical to that of our classical pinball machine. Thus, according to Sect. 1.2, the setup of Fig. 2.1 describes a 1 b device. There also exist partially transparent mirrors with a reflection coefficient different from the transmission coefficient, in which case the distribution of counter responses, while still random, will not be equiprobable; this would correspond to a < 1 bit device (see Fig. 1.2). In the classical pinball machine, the randomness was caused by small fluctuations of the ball's motion as it rolls and hits the pin (or by deterministic chaos right at the pin's edge) – what is the cause for randomness in the case of photons incident on a semireflecting mirror? Well, the cause is not to be found in anything happening before or at the half-way mirror: It is the presence of detectors downstream! We will explain this puzzling fact a little later; for the time being let us thus affirm that it is impossible to predict even in principle which of the counters will respond. In the quantum case, predictions can be made only for the *average* number of counts, i.e., their probabilities. Leaving aside the cause for randomness, the behavior seems to be very similar to our classical pinball machine: There should be only two possible final states of the system. One is "photon into counter D_0," the other is "photon into counter D_1."

But now comes another surprise: If the detectors are taken out of the way and the photons are let undergo further quantum interactions instead of being counted, we find that one single photon behaves as if it was taking both paths *at the same time!* There is no random "either/or" at the beam-splitting mirror B_1. The instant we interpose the counters again and turn them on (one is really enough), even if we do this *after* the time the photon was expected to pass the beam splitter, the photon – the whole photon, not a "half-photon" – appears traveling nicely along one path *or* the other, triggering the counter at the corresponding end! In summary, in a classical pinball machine, after activation, the ball follows a definite path but we just cannot predict which it will be. In a quantum pinball machine the "ball" does not *have* a path until a measurement is made to reveal which path was followed.

Let us illustrate how all this has been demonstrated in the laboratory. The task is to find out how a photon behaves in absence of counters D_0 and D_1. To that effect, we bring both paths together with the help of normal mirrors as shown in Fig. 2.2, make them intersect at point B_2 and only then let them fall into the two counters. If there is nothing at B_2, we have a situation formally identical to that of Fig. 2.1 (except for the two regular mirrors M_1 and M_2), and either D_0 *or* D_1 will respond for each photon injected by the source. Notice carefully that we have reversed the label of the counters so that in absence of both beam splitters we can designate the undisturbed trajectory "$SB_1M_1B_2D_0$" as a "basis path" or *basis state* of the photon, equivalent to the left path in Fig. 2.1. The other basis state would be the path "$B_1M_2B_2D_1$" followed by a photon injected symmetrically from the top left.[†] We will come back to this question of basis states in the next section.

[†] At this stage we also have other choices for the designation of basis states in the system. For instance, we could have chosen as bases "photon on the right side of the median" and "photon on the left side." This is commonly used in the introductory literature on quantum computing. But in the absence of both beam splitters it would require flipping base every time the undisturbed photon crosses the median plane – yet there is nothing there, physically!

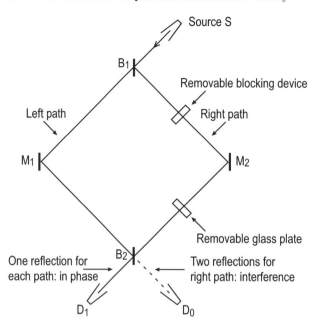

Fig. 2.2. Same as in Fig. 2.1, with two additional regular mirrors M and an additional beam splitter B_2. The path "$SB_1M_1B_2D_0$" represents the undisturbed trajectory of a photon from S in absence of both beam splitters, to be considered as one of the "basis states" in our description of the system. The other basis state is the undisturbed path (via M_2) of a photon incident from the top left side. See text for further detail. With half-silvered mirrors and detectors in place this is a *Mach–Zehnder interferometer*

Now we place a second semireflecting mirror at the crossing point B_2, as shown in Fig. 2.2. Intuitively, there should be no change in the configuration of responses: A photon following the left path between B_1 and B_2[†] should either go straight through B_2 and be counted with counter D_0, or be reflected and trigger counter D_1; a photon traveling along the right side would have the same alternative – only that in this case counter D_0 would register reflected particles whereas D_1 would count transmitted ones. *But this is not what happens!*

To better understand the experiment, let us do it first with a classical coherent, monochromatic light beam. In the absence of both half-silvered mirrors, the light beam would trace what we have called a "basis path" $SB_1M_1B_2D_0$ (a light beam impinging symmetrically from the top left would trace the other basis path). As described above, with the semireflecting mirror in place at B_1, the beam that goes straight through (left path) changes only

---

[†] Henceforth, whenever we say "left path" or "right path," we mean the portion of the path in the section of the device *between* the two beam splitters.

its intensity (to $\frac{1}{2}I$); the reflected beam changes its intensity the same way but also suffers a *phase shift* of $\frac{\pi}{2}$ (in reality, both may suffer phase shifts, but the reflected beam will always lead by $\frac{\pi}{2}$). In the absence of a second beam splitter at B_2, both split beams cross each other and, like in the case of Fig. 2.1, fall into D_0 and D_1, each one of which will receive half the initial intensity. However, when we place a semireflecting mirror at B_2, a double superposition will occur:

1. Detector D_1 will receive half of the right-side beam with unchanged phase (i.e., still leading by $\frac{\pi}{2}$ with respect to the left side), plus half of the left beam reflected with a phase change of $+\frac{\pi}{2}$; therefore both superposed components will be *in phase*, and their intensities will add: The initial intensity I is restored!
2. Detector D_0 will receive *nothing at all:* The right-side beam, when half of it is reflected at B_2, will experience yet another $\frac{\pi}{2}$ phase change, placing it 180° (half a wavelength) *out of phase* with the beam that comes straight through from the left. This means *destructive interference* and zero intensity.

In summary, D_1 will receive the full, original beam of light, whereas D_0 gets nothing.

Note that all this is true only if the lengths or durations of both paths from B_1 to B_2 are exactly equal. If they differ (which can be achieved by inserting a sliver of glass in one of the paths, or an appropriate assembly of mirrors), the relative phase shifts may no longer be the same as described above, and *both* detectors would be illuminated with different intensities I_0 and I_1 (always $I_0 + I_1 = I$). In particular, if the path length differs by odd multiples of half a wavelength (e.g., a phase shift by π in one of the paths), the response will be inverted: Detector D_0 will get all the light and D_1 nothing. If we block one of the paths, both detectors will be equally illuminated, now at 1/4 the initial intensity. On the other hand, if we blow some smoke into the right-side light beam, part of the light will be scattered by aerosol particles; this will allow us to actually *see* that beam, but the interference at B_2 will no longer be complete: Detector D_0 will begin registering some light – up to 1/4 of the initial intensity I_0 if the smoke completely absorbs the right-side beam. The experimental arrangement shown in Fig. 2.2 is called a Mach–Zehnder interferometer.

Now we turn the source intensity down until only one photon is passing through the system at any time. Remember that in this quantum limit, the "intensity of the beam" is related to the *number* of photons per unit time emitted by the source. With no second beam splitter in place we again obtain the situation of Fig. 2.1: The photon will be counted either by D_0 or by D_1. By looking at the result, we obtain *one bit* of information. If we place the second semireflecting mirror at B_2, the experimental result is, of course, entirely consistent with the macroscopic light beam case: *Every* photon emitted by S will be counted by detector D_1 and *no* photon will reach detector D_0. By looking at the result, we obtain *no new* information. In other words, we start with a photon in a known state (a basis state that is an eigenfunction of the "which path" observable). Then, between B_1 and B_2, the state is a superposition of pure states and we cannot know which way it is going. After the second beam splitter at B_2, the photon is again in a known basis state – but it has switched to the other one (which pertains to a photon which enters from the top left when beam splitters are absent).

2.3 The Counterintuitive Behavior of Quantum Systems

The previous statements obviously pose a severe problem of interpretation. We have no difficulty understanding intuitively what happens in Fig. 2.2 with an electromagnetic wave as we turn its intensity gradually down – as long as we view it in our mind as a continuous fluid-like flow of light. Our intuition breaks down the moment we switch to the quantum particle picture at extremely low intensities (relation (2.3b)). Our brain trained in a classical world obliges us to follow mentally each quantum along one of its possible courses between the two beam splitters – but the fact is that as long as the experiment is not arranged to determine the photon's actual path, this particle does *not* follow a single course! We must not imagine the photon as "splitting" in two either, because a split photon has never been observed – we would only encounter a whole one if we were to look for it!

From the informational point of view, notice the following "complementarity." With the arrangement of Fig. 2.2 we do know ahead of time *where* exactly the photon will show up at the end (counter D_1), but we cannot find out which path (right or left) it has followed before getting there – looking at the result of the experiment we obtain no information on this question. This means that there is a region of intrinsic and irremovable uncertainty in the device; yet it leads to a well-defined state at the next step. If we remove the second beam splitter B_2, our measurement will give us one bit of information about *which path* a photon has followed – but we were unable to tell ahead of time where exactly the photon will show up (which counter will respond). The concept of "interference of one photon with itself," and the statement that "one particle seems to follow two different paths at the same time" have no classical equivalents.[†] The intuitively justified question (like in the classical pinball machine) of "what the photon *really* did before reaching B_2 while we were not looking" makes no sense in the quantum domain – because if we

---

[†] Strictly speaking, these statements make little sense even in QM. It is the "full" *wave function*, solution of the time-dependent Schrödinger equation (which in the treatment above we have implicitly averaged out in order to work just with greatly reduced degrees of freedom "left path," "right path" and phase), that actually splits (is nonzero) along both paths and exhibits interference at the second beam splitter.

wanted to find out, we would have to change the experimental setup in such a way that it would invariably spoil the original situation with the interference (see [45])! However frustrating to our intuition, a necessary condition for the interference of a quantum "with itself" is that the experiment be such that it is impossible, even in principle, to obtain information on the particular path the quantum has taken. In other words, quantum systems, while left alone, follow the proverbial policy of "don't ask, don't tell!"

This is so important, not only for the understanding of quantum computing but also for our discussion on information *per se* in Chap. 3, that we should dwell a little more on the setup shown in Fig. 2.2. The following gedankenexperiment will help capture the unexpected features a little better. Suppose we insert a device to block the right-side path at point C – but it is a stochastic device that sometimes works and sometimes does not. Its construction is such that when it does not work, the photon goes through unperturbed; when it does work, the photon is absorbed.[†] We have no control over its function. After the source S fired one photon into the machine, how can we find out if the blocking device has worked or not? We have three alternatives here.

1. This one is trivial: If no count is registered in either D_0 or D_1, we must conclude that the photon was absorbed before reaching B_2 – therefore it has taken the right-side path between the beam splitters, and the blocking device did work.

2. If counter D_0 responds, we also can be certain that the right path was blocked because no D_0 response is possible when both paths are open. So in this case the particle must have followed the left path and gone straight through B_2 into D_0 – but how in the world did it "know" that the blocking device on the right side was working without ever having come in contact with it?

3. If counter D_1 responds, we cannot tell anything about the absorber because if it did work, the photon could have taken the open path on the left side and been reflected at B_2; but if it did not take that route, both paths would have been open and it would have gone into D_1 as the only possibility.

In addition to all this there is a "spooky" temporal aspect. We could insert the "sometimes failing" blocking device *after* the injected photon has passed the first beam splitter, and the experiment would still work (an equivalent operation can be done nowadays in the laboratory). So not only does a photon going through the left side "know" that there is a blocking device in working conditions on the right side – it seems to have known it even before that device was actually inserted or activated! Let us go back to our example in which we blew a cloud of smoke into the classical, continuous light beam on

---

[†] Such a device is easier to design for charged particles, e.g., a sweeping magnetic field, that turns on and off stochastically.

the right side of the device. This example dramatizes the fact that even weak attempts to extract information on which path was followed by a photon will degrade the degree of interference. Knowing that each beam in reality consists of a flux of zillions of photons/$(m^2 \cdot s)$, does all what we said above mean that scattered photons that hit our retina (thus revealing their path) have been in a basis quantum state (the right-side path), not a superposition, because they "knew" right at the first beam splitter that they would be hitting aerosol particles and be scattered? And does it mean that photons registered by the otherwise "forbidden" detector D_0 (thus revealing that they went along the smoke-free left-side path) were in a basis state, too, because they "knew" that, had they been in a superposition, they would have encountered an aerosol particle on the right side? And, finally, does it mean that most of the photons registered by detector D_1 were in a superposition state following both paths at the same time, because they "knew" that nothing would be in the way on their right leg? The answer is: *Yes,* to all of the above!

But hold it! We have been assigning some very anthropomorphic capabilities to the photon like "knowing something" – but a photon does not have a brain to know! We humans do, however, and our classical brain is desperately trying to make classical models of quantum systems in which "information" (about the blocking device, or the aerosol particles) is running around inside the experimental setup at superluminal speed, as well as backwards in time, to intercept the photon wherever it may be, and tell it what to do! The trouble with understanding QM is that we cannot force a description with classical information on a system that is not knowable in classical terms as a matter of principle (see Sect. 5.1). In our present examples all experiments clearly and mercilessly show that if it were possible, or just potentially possible, to extract information about which path the photon has followed, no quantum interference effect would arise. If we could only extract *partial* statistical information about the path, as in our smoke experiment, we would obtain only partial interference. All this is an expression of complementarity, similar to the impossibility of making exact simultaneous measurements of position and momentum. And, evidently, all this should also apply to the case when we just *think* about quantum particles!

Going back once more to our earlier discussion of a beam of light whose intensity is gradually being turned down: Once the quantum limit has been reached *we are prohibited by law* (Heisenberg's uncertainty principle (2.1a)) to visualize mentally the system as a stream of a priori identified and "marked" particles which we then can follow individually as they move through the experimental setup. If we insist, we will indeed mentally see particles follow one path *or* the other like the ball in the classical pinball machine, but *no* interference further down would be possible[†] – that is our punishment for violating

---

[†] It is important to clarify the meaning of "interference" in the case of one photon. Regarding a system like shown in Fig. 2.2, it is sometimes stated that "the photon follows both paths at the same time and then interferes with itself." Again a split

the law! The fact that the case with individual photons is 100% compatible with the macroscopic electromagnetic experiment points to a continuity of *physical* behavior through the classical quantum transition (correspondence principle); what appears as discontinuous in this transition is *our* ability to mentally picture and interpret intuitively the results! We must learn to cope with the fact that in a quantum system there is a clear and irrevocable distinction between its quantum state (while we are not looking – even if the "looking" is done only in our mind!) and the objective properties revealed by a measurement (information delivered to the macroscopic domain). We must realize that a quantum system is both deterministic and unknowable at the same time! Let it be said once and for all: Quantum paradoxes are paradoxes because of our unavoidable urge to represent mentally what happens *inside* the quantum domain with familiar images and paradigms from our sensory experience in a classical, macroscopic world. But nothing that transcends physically or informationally from a quantum system into the macroscopic world is intrinsically paradoxical or contrary to the laws of classical and relativistic physics.

2.4 Basic Algorithms for Single-Particle Quantum Systems

There are many different kinds of experiments with single particles which exhibit features like those discussed in the previous section. For instance, instead of using a semireflecting mirror one can split the paths of photons according to their polarization with a nonlinear birefringent crystal (Sect. 2.6); or use electrons and diffract them in a double slit experiment or split their paths according to their spin in an inhomogeneous magnetic field; or one can use atoms and take two energy levels as the two possible quantum states. These experiments show a fundamental property of all quantum systems, valid as long as the system is left undisturbed (free from irreversible interactions with the outside macroscopic world), namely, the possibility of being in a single state made up of *the superposition of two or more basis states*. By "superposition" we do not mean that the system is sometimes in one, and sometimes in another state: It is *simultaneously* in two or more component states (principle of superposition). We already stated that there is no classical equivalent to this situation. The basis states are eigenstates of an observable – in this case, the "which path" observable – and they are the only possible states in which the system can be found when that observable has actually been measured ("left path" or "right path" in our examples

---

photon has never been seen, and the concept of interference assumes a statistical character in QM: In our example, any deviation from equiprobability (after the *second* beam splitter B_2) represents interference – *total* interference when only one detector responds.

of Fig. 2.1 and Fig 2.2; vertical or horizontal polarization of a photon in a birefringent crystal; spin up or spin down of a particle, etc.).

All this can be expressed in mathematical form using the vector/matrix formalism given in Sect. 1.5. Let us call $|\psi\rangle$ the general state of the photon inside our "quantum pinball machine" after leaving the beam splitter B_1 (Fig. 2.1), and $|\phi_{\text{left}}\rangle$ and $|\phi_{\text{right}}\rangle$ the basis states we have chosen to represent the two possible spatial modes of the system, "photon on left path" (path $SB_1M_1B_2D_0$ in Fig. 2.2) and "photon on right path" between the two beam splitters B_1 and B_2 (please do not get confused with the fact that $|\phi_{\text{left}}\rangle$ pertains to a photon source on the right side and that at the output side it goes into a detector on the right side!). Note that we do not need to consider any continuous spatial variables in $|\psi\rangle$ for our task because we are not interested in where the photon is at any given time (within the limitations of Heisenberg's uncertainty principle), nor are we interested in what exactly the photon does while it is interacting with the semireflecting mirror or any other gadget that we may introduce in the setup. To go into that kind of detail would require integration of the time-dependent Schrödinger equation (Sect. 5.6) with all physical initial and boundary conditions in place, including everything that might interact with the photon as time goes on, like the motion of possible aerosol particles in our smoke example or the "sometimes working" blocking device. In our case, however, the only external degrees of freedom of interest are the binary "which path" observable and, as we shall discuss below, the phase difference; the rest of the degrees of freedom can be "integrated out" and included as a constant factor in the two-dimensional state vector. This procedure of simplification is not unlike the adiabatic descriptions of classical systems, such as a magnetized plasma, in which we deliberately "average out" the rapid cyclotron motion of charged particles about a magnetic field line, replacing each charge by a fictitious particle placed at a "guiding center" and attaching to it an invariant magnetic moment.

The principle of superposition tells us that a general state of the photon between B_1 and B_2 would be:

$$|\psi\rangle = c_{\text{L}}|\phi_{\text{left}}\rangle + c_{\text{R}}|\phi_{\text{right}}\rangle \qquad (2.6a)$$

where c_{L} and c_{R} are two complex numbers.[†]

Why *complex* numbers? For the same reason one uses complex numbers and functions in the mathematical treatment of wave phenomena and electric circuits with alternating current: They come in handy as shorthand to describe a physical quantity that has *both* amplitude and phase. For that reason, in electronics, for instance, voltage and intensity of an alternating current are described by complex functions and the impedance is represented by a

---

[†] $c_{\text{L}}|\phi_{\text{left}}\rangle$ (or $c_{\text{R}}|\phi_{\text{right}}\rangle$) would be the integral of the full-fledged solution of the Schrödinger equation along the left (right) path.

complex number. In QM, given a state that is the superposition of components along basis vectors like (2.6a), one must be able to extract the possible measurement results of a given observable (always real numbers) and their probabilities of occurrence for the state in question (also real numbers); these will depend on the relative phases of the corresponding eigenstate components (see example in relation (2.12)). The need for such an approach should be evident from our discussion of the interference effect regarding Fig. 2.2: Indeed, in the macroscopic case of a light beam, it is the *phase difference* ($\frac{\pi}{2}$) that distinguishes the electromagnetic wave in the left beam from that of the right beam, both of which have equal intensity. We shall retain this macroscopic property as an internal degree of freedom in the quantum description of a single photon in our device (correspondence principle).[†]

In the quantum formalism the values of $c_L c_L^* = |c_L|^2$ and $c_R c_R^* = |c_R|^2$ (the star indicating complex conjugate) are the probabilities of finding the system respectively in state $|\phi_{\text{left}}\rangle$ or $|\phi_{\text{right}}\rangle$ after a measurement was made to find out which path was taken (Fig. 2.1):

$$p_{\text{left}} = |c_L|^2 \quad p_{\text{right}} = |c_R|^2 \,. \tag{2.6b}$$

Since their sum must be $= 1$, we require the normalization condition (2.5c):

$$|c_L|^2 + |c_R|^2 = 1 \,. \tag{2.6c}$$

With this normalization, relation (2.6a) can also be written in polar form $|\psi\rangle = \cos\alpha |\phi_{\text{left}}\rangle + e^{i\varphi} \sin\alpha |\phi_{\text{right}}\rangle$ in which $\cos^2\alpha = p_{\text{left}}$ and $\sin^2\alpha = p_{\text{right}}$. This expression brings out explicitly the phase difference φ. We will come back to this form later.

The physical meaning of the probabilities p_{left} and p_{right} is the same as that of the probabilities p_0 and p_1 in our classical pinball machine (Sect. 1.2): repeating the experiment N times (with $N \to \infty$) under *exactly* the same conditions, $p_{\text{left}} = N_{\text{left}}/N$ and $p_{\text{right}} = N_{\text{right}}/N$, where the enumerators are the number of occurrences of a right path and a left path, respectively. It is important to emphasize, however, that two binary superpositions with the same probability amplitudes (2.6b) but different relative phase φ, do *not* represent the same state. For instance, the symmetric superposition $|\psi_s\rangle = \frac{1}{\sqrt{2}} [|\phi_0\rangle + |\phi_1\rangle]$ (phase difference $= 0$) does *not* represent the same state as the asymmetric superposition $|\psi_a\rangle = \frac{1}{\sqrt{2}} [|\phi_0\rangle - |\phi_1\rangle]$ ($= \frac{1}{\sqrt{2}} [|\phi_0\rangle + e^{i\varphi}|\phi_1\rangle]$, with $\varphi = \pi$), although the probabilities of finding the system in $|\phi_0\rangle$ or $|\phi_1\rangle$ after a measurement are the same. The difference between $|\psi_s\rangle$ and $|\psi_a\rangle$ does not manifest itself statistically in many measurements, but in their behavior when a single quantum transformation is performed, as we shall see in the following sections.

[†] In the usual descriptions of a Mach–Zehnder interferometer, this 90° phase shift it not entered as an explicit variable – only the value of the phase shift caused by an *additional* device inserted as shown in Fig. 2.2 (see also later, relation (2.17)).

Coming back to the quantum pinball machine, after a measurement was made to find out which path the photon has taken, the system would no longer be in a superposition; it would be in a basis state (eigenstate) of the "which path" observable – the particular state that has been identified experimentally. If we use only one counter in Fig. 2.1, say D_0, and it did not register anything, we *know* that the photon has gone through the other path (is in that particular eigenstate – $|c_R|$ would be $= 1$); measuring it again (which is what counter D_1 would do) will give the same result (the information about the path) with certainty.

2.5 Physics of the Mach–Zehnder Interferometer

To describe the operation of the Mach–Zehnder interferometer (Fig. 2.2) quantitatively in the QM case for individual photons, we must find mathematical representations of the basis vectors and the operators responsible for the transformations that the photon state undergoes between the source and the exit. To accomplish that, we must take into account three experimental facts derived from the macroscopic experiment with a continuous coherent light beam:

1. In a symmetric beam splitter there is a 50/50 partition of the intensity (i.e., equal numbers of photons/$(m^2 \cdot s)$ (2.3b)).
2. There is a phase difference of $\frac{\pi}{2}$ between the reflected and transmitted beam at each beam splitter.
3. Any phase change at the fully reflecting mirrors M would be the same for both beams, thus not contributing to a phase *difference* between both beams.

The path lengths are supposed to be equal within tiny fractions of a photon wavelength. For the experiment, we may insert a phase changer between the two beam splitters that introduces an additional phase difference φ between the two channels. We already mentioned that a thin glass plate of appropriate thickness can do the job. To represent the phase variable (an observable) we choose the angle between the polar representations of the complex numbers c_L and c_R in relation (2.6a). For instance, if c_L is real, $\tan \varphi = \operatorname{Im} c_R / \operatorname{Re} c_R$ (see below).

As to the choice of the vectors for the basis states, whenever we have a quantum system that has two and only two possible basis states, we can use the orthogonal unit vectors (1.6), as long as it is made unambiguously clear to which of two, and only two, physical modes (degrees of freedom) each one corresponds. We choose the orthogonal pair

$$|\psi_{\text{left}}\rangle = \begin{pmatrix} 1 \\ 0 \end{pmatrix} \qquad\qquad |\psi_{\text{right}}\rangle = \begin{pmatrix} 0 \\ 1 \end{pmatrix}$$

in which, remember, the first represents the path $SB_1M_1B_2D_0$ in Fig. 2.2 (the "undisturbed path" in absence of both beam splitters) and the second represents the undisturbed path for a photon incident from the left $(B_1M_2B_2D_1)$. At this stage I suggest to any reader less interested in the mathematical treatment to *go directly to relation* (2.11), and take it from there. I must caution, however, that with such a quantum jump he/she may miss several fundamental insights that are important for a better understanding of matters relating to quantum information and its applications.

Initially (between the source S and the first beam splitter B_1), the photon in Fig. 2.1 is in a basis state:

$$|\psi_{\text{initial}}\rangle = |\psi_{\text{left}}\rangle = \begin{pmatrix} 1 \\ 0 \end{pmatrix}. \tag{2.7}$$

We now must describe mathematically the operation of the beam splitter (what we called a "gate" in the classical case, Sect. 1.5) in mathematical, matrix form. In order to conform to the classical case of a light beam, we must find a transformation that converts the initial state (2.7) into a superposition (2.6a) in which $|c_L|^2 = |c_R|^2 = \frac{1}{2}$. Next, we must take into account that there is a phase shift of $\frac{\pi}{2}$ between the right path and the left path [30]; without any loss of generality, we can select $c_L = \frac{1}{\sqrt{2}}$ and $c_R = (\frac{1}{\sqrt{2}})\exp(i\frac{\pi}{2}) = \frac{i}{\sqrt{2}}$ (remember that $\exp(i\frac{\pi}{2}) = i$). Considering the left–right symmetry of the half-silvered mirror in Fig. 2.1, for an incident beam from the *left* side the relations would have to be reversed: The initial state would be $|\psi_{\text{initial}}\rangle = |\psi_{\text{right}}\rangle = \begin{pmatrix} 0 \\ 1 \end{pmatrix}$ and the transmitted beam would emerge on the right side, whereas the reflected beam would appear on the left.

We now must create a matrix $\boldsymbol{S}$ that performs the following symmetry-preserving operations:

$$\boldsymbol{S}\begin{pmatrix} 1 \\ 0 \end{pmatrix} = \frac{1}{\sqrt{2}}\left[\begin{pmatrix} 1 \\ 0 \end{pmatrix} + i\begin{pmatrix} 0 \\ 1 \end{pmatrix}\right] \quad \text{for an incident beam from the right side,}$$

$$\boldsymbol{S}\begin{pmatrix} 0 \\ 1 \end{pmatrix} = \frac{1}{\sqrt{2}}\left[\begin{pmatrix} 0 \\ 1 \end{pmatrix} + i\begin{pmatrix} 1 \\ 0 \end{pmatrix}\right] \quad \text{for an incident beam from the left side.}$$

It is easy to verify that the following matrix fulfils these conditions:

$$\boldsymbol{S} = \frac{1}{\sqrt{2}}\begin{pmatrix} 1 & i \\ i & 1 \end{pmatrix} = \frac{1}{\sqrt{2}}[\boldsymbol{I} + i\boldsymbol{X}]. \tag{2.8}$$

For the last equality we have taken into account relations (1.7a). As a matter of fact, we could have started directly from here: Physically, the half-silvered symmetrical and nonabsorbing mirror performs two operations at the same time: 1. It recreates a component in the input (basis) state with 50% amplitude (hence the factor $\frac{1}{\sqrt{2}}$), and 2. it creates a component in the complementary basis state, also with 50% amplitude and a $\frac{\pi}{2}$ phase shift. The first operation is carried out by the identity operator $\boldsymbol{I}$, the second one by the operator $\mathrm{i}\boldsymbol{X}$.[†] Note several important facts:

1. The operator $\boldsymbol{S}$ does *not* represent an observable but a transformation.
2. $\boldsymbol{S}$ does not create two separate mixed states, it creates two components (projections) of *one* single superposed state in an orthogonal 2D basis system (Hilbert space).
3. $\boldsymbol{S}$ is a *unitary* matrix: $\boldsymbol{SS}^* = \boldsymbol{I}$[‡] (an important property in QM).
4. $\boldsymbol{S}^2 = \mathrm{i}\boldsymbol{X}$.
5. The two superposed states respectively generated by a right and left source input are *orthogonal* to each other.

Turning to our specific case of Fig. 2.1, we again write for the superposed state

$$|\psi_{\mathrm{sup}}\rangle = \boldsymbol{S}|\psi_{\mathrm{initial}}\rangle = \frac{1}{\sqrt{2}} \begin{pmatrix} 1 & \mathrm{i} \\ \mathrm{i} & 1 \end{pmatrix} \begin{pmatrix} 1 \\ 0 \end{pmatrix} = \frac{1}{\sqrt{2}} \left[\begin{pmatrix} 1 \\ 0 \end{pmatrix} + \mathrm{i} \begin{pmatrix} 0 \\ 1 \end{pmatrix} \right]. \qquad (2.9)$$

This is the superposed state of the photon as it exits the half-silvered mirror B_1, i.e., the state of our "quantum pinball machine" of Fig. 2.1 *in absence* of any attempt to measure which path the photon actually took. Let us see what happens if we now subject this superposed state to the *second* beam splitter B_2, as in the device of Fig. 2.2. Applying the gate matrix $\boldsymbol{S}$ again (see relations (1.7a)):

$$|\psi_{\mathrm{final}}\rangle = \boldsymbol{S}|\psi_{\mathrm{sup}}\rangle = \boldsymbol{SS}|\psi_{\mathrm{initial}}\rangle = \mathrm{i}\boldsymbol{X} \begin{pmatrix} 1 \\ 0 \end{pmatrix} = \mathrm{i} \begin{pmatrix} 0 \\ 1 \end{pmatrix}. \qquad (2.10)$$

Since there is no other competing state at this output level, as mentioned earlier, the above phase factor i is physically irrelevant (although it does remind us that, with respect to the initial state, there has been a $\frac{\pi}{2}$ phase shift). Note the fact that the operator $\boldsymbol{S}$ (2.8) has transformed the superposition (2.9) into a single basis state (2.10), which is a flipped, phase-shifted version of the initial state and represents the puzzling fact that under these conditions the outcome of the Mach–Zehnder interferometer is predetermined – a zero Shannon information device. This is a general property: Given a maximally superposed state, there always exists a transformation that can change the system into a single basis state. At this stage there is no way of finding out which path the photon has taken between B_1 and B_2 – as a matter of fact, we have a proof that it must have taken both! Figure 2.3a depicts the situation schematically.

[†] Individual transform operators that act *simultaneously* in one physical device are added; operators that act sequentially representing different devices separated in space and time (such as the second half-silvered mirror, or several phasechangers – see below) are multiplied in the order in which they appear in the time sequence.

[‡] In general, a matrix $\boldsymbol{A}$ of (complex) elements a_{ik} is unitary if $\sum a_{ik}a_{kj}^* = \delta_{ij}$.

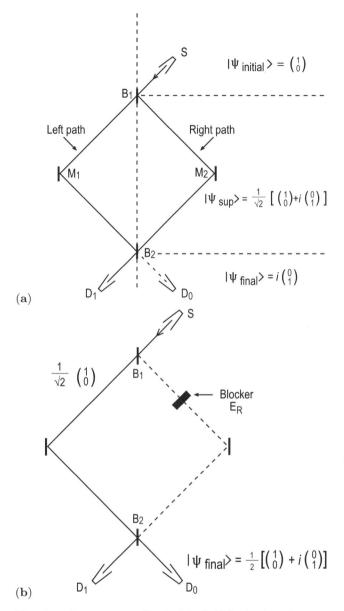

Fig. 2.3. State vectors for the Mach–Zehnder interferometer. (**a**) Operation without modifications of the paths. Do *not* look at the left and right channels as if they would represent independent mutually exclusive paths of the photon! Note where the interference occurs. Since the final state is a priori known, this represents a zero bit device. (**b**) Blocked right path, no interference. The mere *presence* of E_R causes a real, physical split of possibilities for the photon at B_1 (as if the photon knew what is there to come!). If a photon is not absorbed (i.e., it goes through the left path between B_1 and B_2), the final state is maximally indeterminate; i.e., we have a one bit device

Let us now introduce a channel blocker. We can take the following matrices (given in (1.7a), but careful with the left–right designations!) for our purpose:

$$\boldsymbol{E}_\mathrm{R} = \begin{pmatrix} 1 & 0 \\ 0 & 0 \end{pmatrix} \qquad\qquad \boldsymbol{E}_\mathrm{L} = \begin{pmatrix} 0 & 0 \\ 0 & 1 \end{pmatrix}.$$

They represent blocking devices on the left and on the right path, respectively ("eliminate one component but let the other stand" operators). Note that $\boldsymbol{E}$ are transformations, not operators representing an instrument (detector proper). Applying the blocking operators to relation (2.9), we obtain (see Fig. 2.3b):

$$\boldsymbol{E}_\mathrm{R}|\psi_\mathrm{sup}\rangle = \frac{1}{\sqrt{2}} \begin{pmatrix} 1 \\ 0 \end{pmatrix} \qquad\qquad \boldsymbol{E}_\mathrm{L}|\psi_\mathrm{sup}\rangle = \frac{i}{\sqrt{2}} \begin{pmatrix} 0 \\ 1 \end{pmatrix}.$$

We again obtain pure basis states, no superposition.[†] The factors $\frac{1}{\sqrt{2}}$ and $\frac{i}{\sqrt{2}}$ play no physical role in this particular case (although they do remind us that compared to the initial state, the probability of obtaining each state is only $\frac{1}{2}$ and that a left-side blockage leads to an output that is phase-shifted by $\frac{\pi}{2}$). Note that no matter where along the path the blockage is inserted, it affects the superposed state all the way from B$_1$ to B$_2$. If the blocking device is $\boldsymbol{E}_\mathrm{R}$ and the photon takes the left path and goes through the half-way mirror B$_2$ (Fig. 2.3b), we obtain a maximally *superposed* state as we did after B$_1$ when there was no blocking (a 1 b device!). Again, the resulting factor $\frac{1}{2}$ has no absolute meaning, although now it reminds us that the probability of each final branch, compared to the initial state, will be only $\frac{1}{4}$.

We can generalize the scheme of Fig. 2.2 and insert a device between the two beam splitters that shifts the phase between the two channels. Refer to the sketch of Fig. 2.3a and note the three regions with distinct states of the photon: 1. the input region between source and the first beam splitter; 2. the region between the two beam splitters; and 3. the output region. Again, no matter where the phase shifter is inserted, it affects the superposed state all the way from B$_1$ to B$_2$. It is easy to verify that a phase shifter inserted on the right side path is described by a phase shift operator

---

[†] Note that although $\boldsymbol{E}_\mathrm{L}\boldsymbol{E}_\mathrm{R} = \boldsymbol{E}_\mathrm{R}\boldsymbol{E}_\mathrm{L} \equiv 0$ ("cancel everything" operator), their sum is $\boldsymbol{E}_\mathrm{L} + \boldsymbol{E}_\mathrm{R} = \boldsymbol{I}$ (the "leave as is" operator). However, $\boldsymbol{E}_\mathrm{L} + \boldsymbol{E}_\mathrm{R}$ is an "impossible" transformation: You cannot eliminate the left component and let the right one stand and *at the same time* (addition of operators!) eliminate the right component and let the left one stand! This shows how careful one has to be when describing quantum systems: Whenever we do something to one component, we must explicitly say (and represent mathematically) what happens to the others. The null product $\boldsymbol{E}_\mathrm{L}\boldsymbol{E}_\mathrm{R}$ (one operator acting *after* the other) is obvious: Eliminate the right component, *then* eliminate the left one (in this example the order of time is immaterial – both operators commute).

$$\boldsymbol{\Phi}(\varphi) = \begin{pmatrix} 1 & 0 \\ 0 & e^{i\varphi} \end{pmatrix}.$$

Notice that $\boldsymbol{\Phi}(0) = \boldsymbol{I}$ and $\boldsymbol{\Phi}(\pi) = \boldsymbol{Z}$ (see relations (1.7)). Note also that the addition of multiple phase shifters with individual phases $\varphi_1, \varphi_2, \ldots \varphi_n$ is represented by the product operator $\boldsymbol{\Phi}(\varphi_s) = \boldsymbol{\Phi}(\varphi_n) \cdots \boldsymbol{\Phi}(\varphi_2)\boldsymbol{\Phi}(\varphi_1)$, ... which indeed yields $\varphi_s = \varphi_1 + \varphi_2 + \cdots + \varphi_n$, an intuitively expected result. The most general expression of the output state of the system is now, using expression (2.10), $|\psi_{\text{output}}\rangle = \boldsymbol{S\Phi S}|\psi_{\text{input}}\rangle$. This leads to:

$$|\psi_{\text{output}}\rangle = \frac{1}{2}\left[(1 - e^{i\varphi}) \begin{pmatrix} 1 \\ 0 \end{pmatrix} + i\left(1 + e^{i\varphi}\right) \begin{pmatrix} 0 \\ 1 \end{pmatrix} \right]. \tag{2.11}$$

For the benefit of those readers who have skipped the preceding mathematical details, this is the general expression for the output state of a Mach–Zehnder interferometer with a *right-side* input and an extra phase shift device inserted on the right side (no generality is lost by this latter fact – insertion on the left side would only change the output state by a sign; the angle φ can be interpreted as the additional *relative* phase shift of the right path component state with respect to the left path one). The imaginary unit i ($= e^{i\frac{\pi}{2}}$) in the last term comes from the $\frac{\pi}{2}$ phase shift between right and left path component states caused by the first half-silvered mirror (Fig. 2.2). Note that in the literature on quantum information (e.g. [16,69]) this phase factor does not appear in this place because, as mentioned before, it is not treated as an extra degree of freedom; this does not, however, change the physics involved (see also relation (2.17) below).

This general relation shows how an extra change in relative phase φ of the superposed state can switch the output state from a basis that was orthogonal to the initial state (for $\varphi = 0$, our original result for Fig. 2.2) to a basis identical to the initial state ($\varphi = \pi$ – this end result would be the same as having an "undisturbed" beam without any beam splitters in place). For all φ values in between we obtain a superposed state. One has to be careful to remember that φ is the phase difference in the superposed state *before* beam splitter B$_2$; with a little algebra of complex numbers one can show the notable fact that the phase difference between the two components of the output state *after* B$_2$ (2.11) is *independent* of φ and equal to π (which stems from the *two* $\frac{\pi}{2}$ shifts in the beam splitters). The phase shifter shifts the relative phase between B$_1$ and B$_2$, and thus controls probabilities after B$_2$. Indeed, according to (2.11) and (2.6b) the probabilities p_0 and p_1 of detecting the photon in detectors D$_0$ or D$_1$, respectively, as a function of the phase difference is

$$p_0 = \frac{1}{4}\left|1 - e^{i\varphi}\right|^2 = \sin^2\left(\tfrac{\varphi}{2}\right),$$

$$p_1 = \frac{1}{4}\left|1 + e^{i\varphi}\right|^2 = 1 - p_1 = \cos^2\left(\tfrac{\varphi}{2}\right). \tag{2.12}$$

Note that for a phase difference of $\varphi = \frac{\pi}{2}$ or $\frac{3\pi}{2}$ we obtain equiprobable components for the output state (maximum uncertainty, maximum amount of information of one classical bit once a measurement has been made). This is a good example of how the relative phase in a superposed state can control the probabilities of occurrence in the measurement of a quantum system after a unitary transformation, and thus reaffirms the need to work with complex numbers.

Let us summarize the transform operators we have encountered thus far for a quantum system with two basis states:

$$\boldsymbol{I} = \begin{pmatrix} 1 & 0 \\ 0 & 1 \end{pmatrix} \qquad \text{identity,}$$

$$\boldsymbol{X} = \begin{pmatrix} 0 & 1 \\ 1 & 0 \end{pmatrix} \qquad \text{flips the two basis states,}$$

$$\boldsymbol{Z} = \begin{pmatrix} 1 & 0 \\ 0 & -1 \end{pmatrix} \qquad \text{shifts relative phase by } \pi,$$

$$\boldsymbol{Y} = \begin{pmatrix} 0 & -1 \\ 1 & 0 \end{pmatrix} = \boldsymbol{XZ} \qquad \text{phase-shifts by } \pi \text{ and flips,}$$

$$\boldsymbol{E}_{\mathrm{L}} = \begin{pmatrix} 0 & 0 \\ 0 & 1 \end{pmatrix} \quad \boldsymbol{E}_{\mathrm{R}} = \begin{pmatrix} 1 & 0 \\ 0 & 0 \end{pmatrix} \qquad \begin{array}{l}\text{eliminates one of the com-} \\ \text{ponents from a superposition} \\ \text{and lets the other stand,}\end{array}$$

$$\boldsymbol{\Phi}(\varphi) = \begin{pmatrix} e^{-i\frac{\varphi}{2}} & 0 \\ 0 & e^{i\frac{\varphi}{2}} \end{pmatrix} \qquad \begin{array}{l}\text{introduces a relative phase} \\ \text{shift of } \varphi \text{ between right and} \\ \text{left components – written here} \\ \text{in symmetrical form.}\end{array} \qquad (2.13)$$

$\boldsymbol{I}$, $\boldsymbol{X}$, $\boldsymbol{Z}$, and $\boldsymbol{\Phi}$ are *unitary* operators; the $\boldsymbol{E}$ represent irreversible physical actions and do not have an inverse.

Each one of the operators shown in (2.13) represents a specific *physical interaction* with a quantum system. Any such action or transform requires the intervention of a macroscopic device (e.g., a beam splitter or a phase changer) just like the measurement of an observable does, but unitary operations represent reversible actions on the system's state and there is no loss of quantum information about the original state, which can be restored by a reverse transformation. It is important to point out that although the devices used for a transformation of the quantum state are macroscopic the actual interaction mechanism belongs to the quantum domain (e.g., a wave length-thin half-reflecting silver layer, the thin sliver of glass in a phase shifter, a narrow diffraction slit, etc.). Such a device participates in the interaction in a "catalytic" way: It *does not change* after the interaction is over – there is no macroscopic trace left that would represent classical information on the state of the quantum system; as a consequence, in a transformation there is no collapse of the wave function despite the fact that the system has interacted with a classical macroscopic device. A particle detector, on the other hand, is a macroscopic device in which the interaction with the quantum system does cause a lasting change: This represents, indeed, a measurement, and has an irreversible effect on the quantum system (see Sect. 5.5).

All measurements (and actions like E) are irreversible; they cause a change in the state of the system from which there is no return – all information about the original state is wiped out. But there is one exception: if the system already was in one of the corresponding operator's eigenstates. However, as mentioned earlier, in that case the Shannon information is zero (there is only one possibility, no alternative). In summary, unitary transformations produce reversible changes in a quantum system with no classical (i.e., macroscopic) consequences, measurements are irreversible transformations of the system with macroscopic external effects.

2.6 Polarized Photons

To consolidate the understanding of the counterintuitive behavior of simple quantum systems (an oxymoron!), let us consider another example in which the polarization of a photon represents the discrete, binary variable. As in the case of the "quantum pinball machine" of Fig. 2.2, we start with a classical light beam, this time linearly polarized. In a linearly polarized electromagnetic wave the plane formed by the oscillating electric field vector and the direction of propagation (which we shall take along the x-axis of a Cartesian coordinate system) is constant. We choose the angle between the direction of the electric field and the (so far arbitrary) y-axis of that frame of reference as the angle of polarization α. This linearly polarized beam enters an anisotropic crystal whose principal axes of polarization are oriented along the directions of y and z (Fig. 2.4). This birefringent crystal is a *polarizing beam splitter,* dividing the incident beam into two components propagating along different paths, one with vertical polarization (the "ordinary" ray or V-path in the figure) and the other one with horizontal polarization (the "extraordinary" ray or H-path). If the polarization angle is $\alpha = 90°$ (initially a vertical polarization), 100% of the light is transmitted along V; if $\alpha = 0°$, 100% of the intensity is found in the H-path. For an arbitrary polarization angle α, the respective intensities obey Malus' law: $I_V = I_0 \sin^2 \alpha$, $I_H = I_0 \cos^2 \alpha$. This can be explained easily with electromagnetism: The electric field vector E of the wave (whose intensity is proportional to $|E|^2$) is decomposed into two components of magnitudes E_y and E_z, each one of which represents a wave propagating along separate paths in the crystal. In principle, there is no phase change between the two emerging waves, but since the speed of light in an anisotropic crystal is polarization-dependent, there could be a timing delay between the two, a fact that can be compensated with a phase shifter. If we bring the beams together in a "reverse" crystal, they combine (their respective electric field vectors add) to form again one linearly polarized beam with the same polarization angle and intensity as the incident beam.

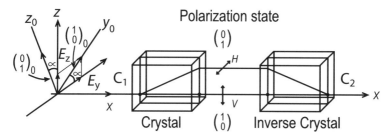

Fig. 2.4. A linearly polarized beam (α is angle of polarization) enters a birefringent crystal and is split into horizontally and vertically polarized beams. In the quantum-level case of one photon, the particle is in a superposed state between C_1 and C_2. The second, inverse, crystal restitutes the original state with polarization angle α

As we did in the Mach–Zehnder interferometer case, we decrease the light intensity until only one photon at a time runs through the setup. Now the polarization must be viewed as an *internal state* of each photon. Let us consider the two paths emerging from the first crystal. Note that in this case there is a direct coupling between which path the photon may take in the crystal and the state of its polarization; as a result, there is no need to introduce a separate, independent, external degree of freedom. From what we have already learned about quantum superpositions, there is no way of knowing which path the photon will follow. If we make a measurement to resolve this uncertainty by placing a detector or absorber along one of the paths between the two crystals (Fig. 2.4), we will find that the V-path will occur with a probability $p_V = \sin^2 \alpha$ and the H-path with $p_H = \cos^2 \alpha$ – but no reconstruction of the actual original state would ever be possible: Any photon "surviving" the measurement or the block emerges with the polarization corresponding to the "unblocked" channel it has taken (V or H). In other words, it will be in an eigenstate of the system (eigenstate of the "which path" observable). Thus, the superposed state between both crystals can be written as

$$|\psi_{\text{sup}}\rangle = c_V \begin{pmatrix} 1 \\ 0 \end{pmatrix} + c_H \begin{pmatrix} 0 \\ 1 \end{pmatrix} \qquad (2.14)$$

in which the orthogonal unit basis vectors represent the two pure states $|\psi_V\rangle$ and $|\psi_H\rangle$, respectively. Assuming that there is no phase factor involved, one can work with real numbers[†] for the coefficients; Malus' law requires that $c_V = \sin \alpha$ and $c_H = \cos \alpha$. Note that a blocked channel converts the birefringent crystal into a *filter:* Whatever the polarization state of the photons fed into it, those which survive the blockage, i.e., have taken the unblocked channel, will always be in a basis state of horizontal or vertical polarization, depending on where the blockage was.

[†] We can add phase shifters and thus introduce states of elliptic polarization, for which we do need complex coefficients.

As we did with the Mach–Zehnder interferometer, we must now find the transformation that links the state of the incoming linearly polarized photon $|\psi_\alpha\rangle$ with (2.14). We have a different situation here. In the example of Fig. 2.2 the beam splitter changes the state of the photon system but the basis vectors remain the same throughout. In the present example of Fig. 2.4 the opposite happens: The state of the photon remains the same when it enters the beam splitter but there is a change of the frame of reference, a rotation. Indeed, when we describe the incoming photon, we can use any pair of base vectors, in particular, the base system belonging to some rotated filter crystal that gave the photon its initial polarization angle α. But in the sector between C_1 and C_2 (Fig. 2.4), the base vectors are fixed, set by the crystal in question. If c_0 and c_1 are the components of the state vector of the photon before entering the crystal (for instance $c_0 = 1$ and $c_1 = 0$, if the unit vector $|0\rangle$ outside the crystal is directed along the $\boldsymbol{E}$ vector – the y_0-axis in the figure), we must find the geometric transformation that links the coefficients in (2.14) with c_0 and c_1.

We can verify that the following matrix $\boldsymbol{R}$ does the job for the pair $c_0 = 1$ and $c_1 = 0$:

$$\boldsymbol{R}\begin{pmatrix} 1 \\ 0 \end{pmatrix} = \begin{pmatrix} \cos\alpha & \sin\alpha \\ -\sin\alpha & \cos\alpha \end{pmatrix} \begin{pmatrix} 0 \\ 1 \end{pmatrix} = \sin\alpha \begin{pmatrix} 1 \\ 0 \end{pmatrix} + \cos\alpha \begin{pmatrix} 0 \\ 1 \end{pmatrix}.$$

It is easy to verify that the inverse of $\boldsymbol{R}$ is a rotation matrix of the same form, with the signs of the sine functions reversed. It represents the second crystal which restitutes the original representation: $\boldsymbol{R}\boldsymbol{R}^{-1} = 1$.

In Fig. 2.4 we have a situation to a certain extent similar to what happens in the Mach–Zehnder interferometer (Fig. 2.2): From being represented by a basis state initially, the photon has gone through a state of full uncertainty (concerning the two possible polarizations and paths between the two crystals), back to a basis state with complete certainty. If we want to resolve the intermediate uncertainty, we must make a measurement, but in that case, the original state cannot be reconstituted. The difference with the system in Fig. 2.2 is that in the Mach–Zehnder case the initial basis state (e.g., right-side input) has been flipped (left-side output).

It is instructive to "play" with polarization filter arrangements, using them first with classical light beams (continuous electromagnetic waves), and then turning the intensity down to just one photon at a time. For instance, a vertically polarized photon will be absorbed in a horizontal filter – but if we interpose a 45° filter, the photon will pass the horizontal filter in half of the cases (this experiment can be made in the classical domain by using three polaroid sunglass lenses!). Now consider this train of thought: Do the photons coming out of the double crystal arrangement of Fig. 2.4 not have *either* horizontal or vertical polarization? If we insert a vertical filter at the end, indeed, we will see $100\sin^2\alpha$ percent of them getting through; the rest will be absorbed. If the filter is horizontal, $100\cos^2\alpha$ percent will pass. But if we insert a filter turned an angle α, *all* of them will pass (see [45])! How do the photons coming out of the second, inverse crystal "know" beforehand that they will encounter a filter turned 90°, 0° or α degrees? This is the best proof that in our arrangement photons are not *either* in one *or* in the other state of polarization, but in a single linear polarization state at angle α. It is the integral experimental setup that determines their behavior, which to us seems as if they were *at the same time* in vertical and horizontal polarization inside the crystal. The presence of a horizontal or vertical filter after C_2 will collapse the photon's state into one of the two eigenstates in the crystals, and we will learn which path it has taken (just as the presence of detectors in the arrangement of Fig. 2.1 causes a collapse into a left-path or right-path eigenstate). But because of the initial conditions, the photon's state is an eigenstate of a filter *rotated* by the polarization angle α; the insertion of such a filter will leave the photon's state untouched and the particle will pass through unscathed.

2.7 Quantum Bits, Quantum Information and Quantum Computing

It is time to turn from QM to *quantum information*. In Sect. 1.5 we defined the classical bit, or Cbit, as a physical device with two well-defined, stable and mutually exclusive states. The system is prepared by some action that causes it to assume one of those states; to find out which one it is, a measurement must be made. The result of the measurement will provide one bit of new information, the maximum amount of information that can be expected from the device when both states are equiprobable. In no way is the measurement expected to perturb that which is being measured (the "which state" binary variable), and there is stability: After any action on the Cbit has ceased, it will remain in the same state until reset or changed by another action. This is a basic property of any binary register in a computer. In Sect. 1.5 we represented the two possible, mutually exclusive, states in which a Cbit can be found by two orthogonal unit vectors in Hilbert space, although this is not particularly useful if one remains in the classical realm – indeed, we did this only to provide elements of reference for comparison with the quantum case.

It is possible to construct binary devices at the quantum level similar to a Cbit. Our "quantum pinball machine," the idealized Mach–Zehnder interferometer of Fig. 2.2, is an example. Consider the photon on its way between the two beam splitters in absence of any detector or obstacle that would allow identification of the particular path taken. We are in presence of what is called a quantum bit, or *Qbit* [76, 100]: a physical system with two basis states (eigenstates of the operator representing the measurement of a binary observable), but *without* the requirement of mutual exclusivity (forget for a moment the extraordinarily short lifetime and the general fragility of this quantum system in our example).[†] In other words, a Qbit is a register which can be prepared in a *superposition* of two different component states – representing an infinite number of possibilities with different relative amplitude and phase. To specify the superposed state of a Qbit one must define two complex numbers (2.6a) bound only by the normalization condition (2.6c). Because of the latter, only two real numbers are required for the specification. For this reason, it is advantageous to use the polar expression mentioned in connection with relation (2.6a) for the state of a Qbit in order to put in evidence the phase:

$$|\psi\rangle = \cos\alpha|0\rangle + e^{i\varphi}\sin\alpha|1\rangle. \qquad (2.15)$$

The normalization condition (2.6c) is now automatically fulfilled. This expression, with the two real parameters α and φ, shows how much information can be encoded at once in just one Qbit! When $\alpha = \frac{\pi}{4}$, we have *maximally superposed* states. A measurement of a system in such a state provides one classical bit of Shannon information (maximum uncertainty *before* the measurement); for any other value of α, the information obtained is always less than one bit. Note that according to expression (2.15) and relation (1.3), the amount of directly extractable information from a Qbit is independent of its relative phase. Note carefully that the phase φ does not affect the probabilities $p_0 = \cos^2\alpha$ and $p_1 = \sin^2\alpha$, i.e., it does not affect the distribution of results if measurements were to be made on many identical Qbits. However, both α *and* φ determine the state of the Qbit; in other words, two Qbit states are different if their phases are different, even if the amplitudes or respective probabilities (2.15) are the same. Going back to relation (2.11) and Fig. 2.3b, the phase difference in state $|\psi_{\text{sup}}\rangle$ would not affect the probabilities if a measurement of the "which-path" observable would be made in that pre-beam splitter 2 stage. But, as shown in relation (2.11), if left untouched (unobserved) it does affect the next stage $|\psi_{\text{output}}\rangle$.

Unfortunately, all this richness in possibilities for the state of a Qbit (the values of α and φ) collapses the instant a measurement is made: The internal state of the Qbit jumps from wherever it was in Hilbert space to point P or Q. With the measurement process we have forced the Qbit to interact with the macroscopic classical world which causes its state vector to collapse onto one of its basis states. The bottom line is only the maximum of *one bit* of classical, i.e. usable, information can be extracted from a Qbit! In general, if there are no further actions on a Qbit, it will remain in the same superposed state until changed by a reversible transformation, by decoher-

---

[†] In general, a Qbit can consist of just one quantum particle prepared and "maintained" in very special conditions. In general, external states like path and position are far more difficult to maintain and shield from outside perturbations by electromagnetic and gravitational fields (decoherence, Sect. 5.6) than internal states like the spin of a particle or the stable energy states of an atom.

ence[†] or by a *measurement*. This latter fact is in contrast to what happens with a Cbit, whose state is immune to measurements. A Cbit holds only one value represented by the state it is in; a Qbit in superposed state assumes a unique value only *after* a measurement. One may call the parameters α and φ in (2.15) or the amplitudes c_L, c_R in (2.6a) the *quantum information* in an algorithmic sense (Sect. 1.4) carried by the Qbit; it would represent a huge amount of classical information (embedded in the real numbers α and φ) if we *could* extract it from the Qbit – but we cannot! Thus defined, quantum information is conferred to the Qbit in the process of its preparation; it can be changed, transferred and made to operate on other Qbits – but it cannot be converted into an equivalent amount of classical, extractable information in a "one-shot" operation (such information can only be extracted statistically, from a large sample of Qbits prepared in identical ways).

Suppose you are handed two little look-alike boxes and are told that one contains a Cbit, the other one a Qbit. You are not allowed to look inside to see their internal structure – all you can do is make a one specific measurement on each to determine the state it is in. Can you find out which device is a Cbit and which is a Qbit? *No!* If you are given hundred samples of each (assuming that each device of the same class has undergone exactly the same history of preparation), you could determine statistically the probabilities of occurrence of their basis states when measured – but you still would not be able to determine which class consisted of Cbits and which one of Qbits, because a "measured Qbit" behaves like a Cbit! Another example, just to emphasize the nature of quantum behavior is the following. You are given 100 Qbits and perform a measurement on each one. You find that half of them are in state $|0\rangle$ and the other half in state $|1\rangle$. Does this mean that all Qbits were in a superposed state of the type (2.15) with $\alpha = \frac{\pi}{4}$ or $\frac{3\pi}{4}$? *No*, because the set could have consisted of a statistical 50/50 mixture of Qbits in a pure basis state $|0\rangle$ or $|1\rangle$. A Qbit in a superposed state is not the same as a Qbit that is *either* in state $|0\rangle$ *or* in state $|1\rangle$, even if statistical measurement results would be the same. (By the way, the phase in the superposed state of a Qbit does not affect the result of a measurement – it affects what the Qbit will do to other Qbits, as long as we do not look into what happens inside!)

There is one other important related restriction. The state of a Qbit *cannot be copied* – this is sometimes called the "no-cloning" theorem – first proposed by *Ghirardi* [44]. Suppose we have two Qbits, one in some basis state $|0\rangle$ and the other in a generic superposed state $|\psi\rangle$, whose history (preparation)

[†] As mentioned before, decoherence is due to an irreversible and, in most man-made situations, unavoidable interaction with the environment. Basically, decoherence destroys the phase relationship between the two component states of a Qbit (2.15), converting it into a classical binary unit (like the classical pinball machine). But in certain experimental situations the effects of decoherence can be made reversible ("erased"). We will come back to decoherence in Chap. 5; at this stage we shall consider it as mere "noise."

is unknown. We want to put them in mutual (quantum) interaction in such a way as to transform the standard Qbit $|0\rangle$ into $|\psi\rangle$, thus obtaining two Qbits in an identical superposed state. This, however, is prohibited by quantum law – indeed, if such process existed, we could make many exact copies of the original state $|\psi\rangle$ and determine statistically, as accurately as we wished, the probability distribution that *defines* that particular state, by making a measurement on each one of those copies. Physically this would mean that we have determined the state of the *original* Qbit without having subjected it to any measurement at all (i.e., to an interaction with a macroscopic measurement apparatus, which would have destroyed its superposed state). The result of this proof *ad absurdum* is that quantum information cannot be copied! This strengthens our initial statement that there is *no process* that can reveal any information about the state of a quantum system without disturbing it irrevocably.

Quantum uncertainty relates to the intrinsic *inaccessibility* of quantum information: I have to disturb the Qbit with a measurement to get anything out at all, and then, the most I can expect to obtain is *one* lousy bit of information. So what is the point of using Qbits, if, once interrogated, they just behave like Cbits? The only way to put the internal "richness" of information of a Qbit to use is to have it evolve through interactions with other Qbits at the quantum level *without looking* into what is happening while the process is going on. The trick is to connect Qbits with each other in a network and let quantum information processing proceed by performing reversible operations (unitary transformations) in some clever way before any classical information is extracted from a final state through measurement. The initial state of such a network is usually a set of Qbits in known basis states $|0\rangle$ or $|1\rangle$ as the result of some classical, macroscopic intervention (by humans or human-programmed devices), such as measurements or filters (see preceding section). During the interactions at the quantum level, appropriately shielded from any classical macroscopic interference, the combined quantum system can evolve in extraordinarily complex fashion and be transformed in such ways as to yield, upon measurement at some "final" stage, comparatively few but significant bits of classical information (providing, for instance, in an extremely short time the prime factors of a large number; or encrypting and transmitting a secret message in an unbreakable way; e.g., [16]).

Let us remember again: There are measurements and there are transformations; both act on a quantum system and both are represented mathematically by operators. Measurements are a class of irreversible transformations; they provide classical information but change the state irreversibly in such a way that it is impossible to find out what it was originally, let alone to reproduce it. Measurements are used to extract output information from a Qbit network. Transformations change the state of the system and with it, the (inscrutable) quantum information it contains in a deterministic way; if the corresponding operator is unitary, the transformation is reversible and the system remains at the quantum level. Such reversible transformations do not yield any classical information – they are gates that change the quantum state.

All transformations are the result of the action on the Qbit of some device such as a half-silvered mirror, a thin glass plate, a birefringent crystal, etc. Any such action is represented mathematically by an operator. In addition to the operators already mentioned (2.13), there is another important operator, which transforms any pure basis state into the superposition of two basis states *without* any mutual phase shift. It is the *Hadamard operator*, a veritable "workhorse" of quantum computing, defined by

$$\boldsymbol{H} = \frac{1}{\sqrt{2}} \begin{pmatrix} 1 & 1 \\ 1 & -1 \end{pmatrix} = \frac{1}{\sqrt{2}} (\boldsymbol{X} + \boldsymbol{Z}). \tag{2.16}$$

It is a unitary operator. $\boldsymbol{H}$ can be used to represent the action of a beam splitter which does not cause any phase changes between the outgoing paths. Operating on right-side and left-side inputs like in Fig. 2.2, respectively, we obtain the following superpositions:

$$|\psi_{\text{sup}}\rangle^+ = \boldsymbol{H} \begin{pmatrix} 1 \\ 0 \end{pmatrix} = \frac{1}{\sqrt{2}} \left[\begin{pmatrix} 1 \\ 0 \end{pmatrix} + \begin{pmatrix} 0 \\ 1 \end{pmatrix} \right]$$

and

$$|\psi_{\text{sup}}\rangle^- = \boldsymbol{H} \begin{pmatrix} 0 \\ 1 \end{pmatrix} = \frac{1}{\sqrt{2}} \left[\begin{pmatrix} 1 \\ 0 \end{pmatrix} - \begin{pmatrix} 0 \\ 1 \end{pmatrix} \right].$$

Both are maximally superposed states. Note the antisymmetry of the second state; the phase difference between its two components is π. A second operation of $\boldsymbol{H}$ on the superposed states returns the original input state *exactly*, without any phase factor (e.g., right side in, right side out). If we apply $\boldsymbol{H}$ to a superposition with a phase shift φ, we obtain

$$\boldsymbol{H}|\psi_{\text{sup}}\rangle = \boldsymbol{H} \frac{1}{\sqrt{2}} \left[\begin{pmatrix} 1 \\ 0 \end{pmatrix} + \mathrm{e}^{\mathrm{i}\varphi} \begin{pmatrix} 0 \\ 1 \end{pmatrix} \right]$$

$$= \frac{1}{2} \left[(1 + \mathrm{e}^{\mathrm{i}\varphi}) \begin{pmatrix} 1 \\ 0 \end{pmatrix} + (1 - \mathrm{e}^{\mathrm{i}\varphi}) \begin{pmatrix} 0 \\ 1 \end{pmatrix} \right], \tag{2.17}$$

an expression formally equivalent to (2.11), but without a factor i in the second term. The fact that the operation $\boldsymbol{H}|\psi_{\text{sup}}\rangle$ returns different states depending on φ in the superposed state, demonstrates again that phase is an important physical quantity! Taking into account (2.13) and (2.16), note the relations $\boldsymbol{XZ} = -\boldsymbol{ZX}$, $\boldsymbol{HX} = \boldsymbol{ZX}$, $\boldsymbol{HXH} = \boldsymbol{Z}$ and $\boldsymbol{HZH} = \boldsymbol{X}$. The latter shows that by inserting a $\varphi = \pi$ phase shifter ($\boldsymbol{Z}$) between two Hadamard operators, we obtain a flip operation (or, taking into account (2.10), $\boldsymbol{HZH} = \boldsymbol{X} = -\mathrm{i}\boldsymbol{S}^2$). These are important relations for the development of quantum computing systems [76].

Although only very small amounts of information can be obtained from Qbits through measurements, huge amounts of hidden quantum information could be processed in parallel between input and output. We can trace an analogy with how the brain functions (Chap. 6): If I am playing a game of chess and my opponent makes a move (the "input," encoded in just a few numbers or symbols), my brain executes a fantastic amount of operations before I make the next move (the "output," also encoded in a few numbers). Neither I nor my opponent know in detail, neuron by neuron, what goes on in my brain, and trying to find it out from the information-processing point of view would probably destroy its state. We both have access only to very limited macroscopic output information. Careful, however, with this brain metaphor: After all, the brain (and any present-day chess-playing computer) is a classical information-processing device with no intrinsic, only practical, limitations to finding out what goes on inside! (As we shall see in Sect. 6.1, the brain does not behave like a coherent quantum system – there are, however, several proponents of this idea (e.g., [85])).

The amount of information a quantum computer could handle is awesomely greater than a classical counterpart with the same number of elements. The state of one Cbit is specified by one bit (Sect. 1.5) – one Qbit instead requires *two real numbers* (2.15)! A classical register of n Cbits can store, say if $n = 3$, *one* integer number from 0 to 7 (the $2^3 = 8$ binary numbers 000, 001, ... 111) at any time. But n Qbits, because of the possibility of superposed and entangled states, can carry *at the same time* the values of 2^n *complex* numbers (subjected to the normalization condition (2.5c)), which would be available for processing with unitary operations. In principle, there is an expansion from a linear capacity in classical computers to an exponential one in quantum computers. A register with 50 Qbits (say, made of 50 atoms, each one with just two possible spin or energy eigenstates) could potentially store and, if acted upon by appropriately chosen unitary transformations, handle up to $2 \times 2^{50} = 2.25 \times 10^{15}$ real numbers in massive parallel processing! However, after such a deluge of operations, in the end only 50 b of classical information could maximally be extracted from the system!

2.8 Entanglement and Quantum Information

So far we have only considered individual Qbits with single particles. The game of quantum computing gets really interesting and promising when superposed states involving more than one particle come into play – but it will also deliver more "collateral damage" to our intuition!

Let us consider two Qbits A and B (carried by two particles of the same kind, or any other appropriate quantum registers). As long as the Qbits are independent of each other (no common origin, no mutual interaction), each one will be represented by a separate state vector of the type (2.15), which we will rewrite now in the form:

$$|\chi_A\rangle = a_A|0\rangle_A + b_A|1\rangle_A \quad \text{and} \quad |\chi_B\rangle = a_B|0\rangle_B + b_B|1\rangle_B\,.$$

The subindices A and B refer to two physical entities (the Qbits) and the vectors $|0\rangle$ and $|1\rangle$ refer to their basis states (in the case of a pair of particles, to some binary internal variable like spin, polarization, pair of energy levels, etc.). Each pair of coefficients satisfies the normalization condition (2.6c). If there is any chance for mutual interaction, whether past, present or future, QM demands that both Qbits be treated as components of *one single system*. In that case, we can use the mathematical formalism for two Cbits given in Sect. 1.5. If each Qbit is in a basis state $|0\rangle$ or $|1\rangle$, there are four possibilities for the combined state of the two-Qbit system shown in relations (1.8). If on the other hand each Qbit is in a superposed state, we can write for the combined state:

$$\begin{aligned}|\chi_{A|B}\rangle &= |\chi_A\rangle|\chi_B\rangle = [a_A|0\rangle_A + b_A|1\rangle_A]\,[a_B|0\rangle_B + b_B|1\rangle_B]\\ &= a_A a_B|0\rangle_A|0\rangle_B + a_A b_B|0\rangle_A|1\rangle_B + b_A a_B|1\rangle_A|0\rangle_B + b_A b_B|1\rangle_A|1\rangle_B\,.\end{aligned}$$

$$(2.18)$$

But beware of the fact that this is *not* the most general superposed state of a two-Qbit system! Such a general state $|\chi_{AB}\rangle$ (notice the slight change in the subindex) is given by the most general linear superposition of the basis states $|0\rangle_A|0\rangle_B$, $|0\rangle_A|1\rangle_B, \ldots$ etc.:

$$|\chi_{AB}\rangle = c_{00}|0\rangle_A|0\rangle_B + c_{01}|0\rangle_A|1\rangle_B + c_{10}|1\rangle_A|0\rangle_B + c_{11}|1\rangle_A|1\rangle_B \qquad (2.19)$$

in which the coefficients are any complex numbers subjected to the normalization condition $\sum |c_{ik}|^2 = 1$, where $i, k = 0$ or 1 (see point P in Fig. 1.5). Only if we impose the additional restriction $c_{00}c_{11} = c_{01}c_{10}$, do we obtain an expression that can be factorized into two separate states like in (2.18), representing two *independent* Qbits, each one in its own superposed state. In the latter case, any operator acting on one of the Qbits (for instance, a transformation or a measurement) will leave the superposed state of the other intact. When $c_{00}c_{11} \neq c_{01}c_{10}$, however, the two Qbits are *entangled*. No longer are their states independent: Manipulating one particle will change the quantum properties of the other – no matter where in space and time the other particle is located! In particular, measuring one Qbit will collapse the *combined* state $|\chi_{AB}\rangle$; therefore, such measurement will also irreversibly change the other Qbit – again, no matter where in space and time it is located!

The measurement of a 2-Qbit system in a state (2.19) will furnish four possible results (pairs of eigenstates and corresponding eigenvalues): $|0\rangle_A|0\rangle_B$, $|0\rangle_A|1\rangle_B$, $|1\rangle_A|0\rangle_B$ or $|1\rangle_A|1\rangle_B$. The probability p_{ik} for each one will be given by the coefficients in (2.19): $p_{ik} = |c_{ik}|^2$. If all coefficients are equal (which means that the condition of independence is automatically fulfilled), a maximum of 2 b of classical information can be obtained (refer to Fig. 1.3a). When the pair of Qbits is entangled, this may no longer be the case.

Just as we can prepare a maximally superposed state for one Qbit, we can obtain maximally entangled states for two Qbits. There are four possibilities of maximum entanglement, called *Bell states*, corresponding to the following conditions for the coefficients c_{ik} in (2.19) (we shall follow closely the description given by *Josza* [61]):

For $|c_{01}|^2 + |c_{10}|^2 = 1$ (which means that $c_{00} = c_{11} \equiv 0$):

$$|\Psi^-\rangle = \frac{1}{\sqrt{2}}\left(|0\rangle_A|1\rangle_B - |1\rangle_A|0\rangle_B\right)$$

$$|\Psi^+\rangle = \frac{1}{\sqrt{2}}\left(|0\rangle_A|1\rangle_B + |1\rangle_A|0\rangle_B\right). \tag{2.20a}$$

For $|c_{00}|^2 + |c_{11}|^2 = 1$ (which means that $c_{01} = c_{10} \equiv 0$):

$$|\Phi^-\rangle = \frac{1}{\sqrt{2}}\left(|0\rangle_A|0\rangle_B - |1\rangle_A|1\rangle_B\right)$$

$$|\Phi^+\rangle = \frac{1}{\sqrt{2}}\left(|0\rangle_A|0\rangle_B + |1\rangle_A|1\rangle_B\right). \tag{2.20b}$$

These four Bell states are also unit vectors normal to each other; they represent the *Bell basis* system in the four-dimensional Hilbert space (see for instance point Q in Fig. 1.5). This means that any general two-Qbit state can be expressed in terms of its projections onto the Bell system. In other words, with an appropriate change of the basis of representation (a rotation in Hilbert space), any general state (2.19) can be expressed as

$$|\chi_{AB}\rangle = C_{11}|\psi^-\rangle + C_{12}|\psi^+\rangle + C_{21}|\Phi^-\rangle + C_{22}|\Phi^+\rangle$$

with the complex coefficients C_{pq} linear functions of the original c_{ik}. As an example, we can express the nonentangled state $|0\rangle_A|0\rangle_B$ in the form $|0\rangle_A|0\rangle_B = (\frac{1}{\sqrt{2}})[|\Phi^-\rangle + |\Phi^+\rangle]$ (note that this is a unit vector at 45° in the $|\Phi^-\rangle$, $|\Phi^+\rangle$ coordinate plane). Considering the unitary operators I, X, Y and Z defined in relations (2.13) and assuming that they act on only *one* of the two Qbits, say particle A, it is easy to verify that all Bell states can be obtained by the following transformations of the asymmetric Bell state $|\psi^-\rangle$:

$$|\psi^-\rangle = I_A|\psi^-\rangle \qquad |\psi^+\rangle = Z_A|\psi^-\rangle$$
$$|\varphi^-\rangle = -X_A|\psi^-\rangle \qquad |\varphi^+\rangle = Y_A|\psi^-\rangle. \tag{2.21}$$

The subindex A reminds us that the operators act on only one of the two Qbits.

What does all this mean, practically? First, let us consider the example of an entangled pair of electrons created in one common process and emitted in opposite directions (particle A on left path, particle B on right path). We impose some conditions on the intrinsic degrees of freedom, such as the spin: If the spin of A is "up" (state $|0\rangle_A$), the spin of B must be "down" (state $|1\rangle_B$), and vice versa. Clearly then, the initial state of the pair will be one of the two Bell states (2.20a) – we shall assume it is the asymmetric one, $|\psi^-\rangle$. A measurement of the spin of A will destroy the superposition and leave the system in a well-defined state, either $|0\rangle_A|1\rangle_B$ or $|1\rangle_A|0\rangle_B$, with a 50% a priori chance for each; after the measurement of the spin of A, the uncertainty about the spin of B will have disappeared – without having performed any actual measurement on particle B itself. This means that despite having two Qbits involved, this Bell state system carries only *one* bit of retrievable information; if A is measured, particle B will carry zero classical information.

It is important to understand this process by comparing it with the classical case. Let us go back to the good old classical pinball machine of Fig. 1.1. This time we place *two* balls in the top bin, a white one and a black one. If we activate the machine with total symmetry, there will be four possible end states, $|0\rangle_W|0\rangle_B$, $|0\rangle_W|1\rangle_B$, $|1\rangle_W|0\rangle_B$ and $|1\rangle_W|1\rangle_B$. A "measurement" of the system will provide 2 b of information. Likewise, two independent Qbits in symmetric or antisymmetric superposed states each would provide the same amount of classical information, once measured. Now we modify our pinball machine mechanically in such a way that two balls cannot fall into the same bin. The only final states allowed will be $|0\rangle_W|1\rangle_B$ or $|1\rangle_W|0\rangle_B$; a measurement now would furnish only 1 b of information (we would always know the color of the ball in the second bin, once we looked into the first one). The result is the same if we had two Qbits in an entangled Bell state – indeed, we may well say that in this modified pinball machine the balls are "classically entangled"! But what is the difference? In the classical case, we know that the balls have been in the bins in which they are found from the very beginning; the measurement process did nothing to the system – only to *our* knowledge of it! In the quantum case, the orientation of the spins of the electrons is *intrinsically* and physically undefined unless a measurement is made; if there is no classical intervention, the system remains in a combined state in which the superposed state of one particle is entangled with that of the other – even if both are far from each other.

There are different experimental ways in which two Qbits can be prepared in a given Bell state. Examples are entangled nuclear spins in nuclear magnetic resonance experiments; parametric down-conversion in which a UV pulse creates a polarization-entangled pair of photons of half the original frequency; and so on. Likewise, there are various ways the Bell state of two Qbits can be identified experimentally (this is called a "Bell state measurement"). It would be far beyond the scope of this book to describe these laboratory methods in detail. Let me just say that this task can be accomplished with beam splitters and, if necessary, phase shifters and/or polarizing filters. To give an example: If two entangled particles in one of the Bell basis states (2.19) arrive simultaneously in the two input channels of a beam splitter (e.g., see Fig. 2.1), the output state will depend on which Bell state (or Bell state superposition) the pair was in. A pair of electrons, for instance, would emerge in separate channels (right path and left path, Fig. 2.1) when they are in the asymmetric Bell state $|\psi^-\rangle$ initially;[†] entangled pairs of photons and other Bosons in the other three symmetric Bell states, would emerge together in a common channel (right path or left path). Those cases in turn could be further discriminated with appropriate filters.

Although the process of measurement in the case of quantum entanglement looks like an instant nonlocal interaction between two different particles which could be located very far from each other (a fact which greatly worried none other than *Einstein* [35]) – there is no violation of causality and theory of relativity: In no way can one send *classical* information instantaneously to where particle B is located by manipulating particle A [46]. An observer (usually called "Bob" in the literature) who holds particle B would have to make an independent measurement to determine its spin, or wait until a message with the pertinent information arrives by classical means from the observer at A (usually called "Alice") – and this takes time! The puzzling thing is, of course, the fact that Bob's measurement result will be consistent with the prediction made by Alice, no matter *when* he makes his measurement – even if he makes it *before* Alice does! Another way of saying: Quantum information *does not propagate,* quantum information *is* – it pertains to the system as a whole, like the wave function itself (Sect. 2.1). Quantum information is not subjected to the classical laws of causality, but then ... it can never be transferred to the macroscopic domain! In general, the collapse of a state function – which *is* instantaneous and nonlocal – *cannot* be used in any way to transmit classical information (extractable information, see below). In more specific terms, the collapse cannot be used to cause an intentional macroscopic *change* somewhere else that otherwise would not occur or would only happen stochastically (the very essence of the concept of information, as we shall discuss in detail in the next chapter). Figure 2.5 illustrates the situation.

A brief interlude is in order concerning the "propagation" of information. Classical information theory does not concern itself with the speed of propagation; it does not prohibit infinite speed – what happens is that when we say "information" in physics we usually mean more than just its measure (which is the only concept with which Shannon information theory deals). Indeed, as we shall see in Chap. 3, there is more to a truly objective concept of information than just its amount: What counts is what information *does!* And it is here where causality and the finite speed of propagation come in. For instance, consider the intersection point of a ruler held at a small angle with respect to a straight line. If I move the ruler in a direction perpendic-

---

[†] Electrons are Fermions (half-integer spin particles) – maximally entangled Fermions can only be in an asymmetric state. The other class of particles, Bosons (integer-spin particles like photons or hydrogen atoms), have to be in symmetric states. A QM principle says that there are only two types of particles: Fermions or Bosons (e.g., [11]).

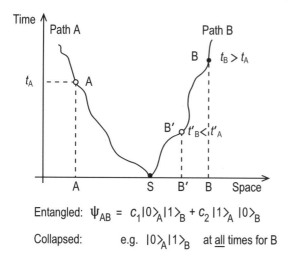

Entangled: $\psi_{AB} = c_1 |0\rangle_A |1\rangle_B + c_2 |1\rangle_A |0\rangle_B$

Collapsed: e.g. $|0\rangle_A |1\rangle_B$ at <u>all</u> times for B

Fig. 2.5. World lines of a pair of entangled particles emitted in opposite directions from point S. At time t_A a measurement is made at point A, which, say, yields for particle A the basis state $|0\rangle$. This means that the complete state of the system has collapsed into the base $|0\rangle_A |1\rangle_B$, and that a measurement of the particle at B will reveal with certitude state $|1\rangle$. Even if the measurement of particle B is made earlier, at time t'_B, particle B still would turn out to be in state $|1\rangle$. So when *exactly* does the state function of the system collapse? See text and Sect. 5.6

ular to the straight line and if the angle is sufficiently small, I can, with my own hands, make that intersection point move with arbitrarily large speed – many times that of the velocity of light! This is another, in this case classical, example of superluminal propagation of a pattern that does not violate relativity in any way, because with that intersection point I cannot transmit "objectively defined" classical information (i.e., transmit a signal with a purpose such as causing a change at some other distant point).

Keeping this argument in mind, let us examine the question: What is, in essence, quantum information? It emerges from *our* mental image – a model – of how certain quantum systems, which *we* have designed for a purpose, work. In the simplest case of one Qbit we represent it by two real numbers (e.g., see relation (2.15)) – two "full" real numbers, not truncated by some experimental error, each one of which may carry a nondenumerable *infinite amount of information*. But to conform to experimental results, this quantum information must be pictured as a nonlocal concept, traveling with *infinite speed* from one point to another, even backwards in time (see discussion of Fig. 2.5). This is, indeed, a double whammy for our classical intuition! But not to worry: We cannot access quantum information directly to verify its strange behavior – we can only access its effects on the classical macroscopic world, and those behave very tamely, indeed! There is no infinite amount of information and no infinite speed of propagation.

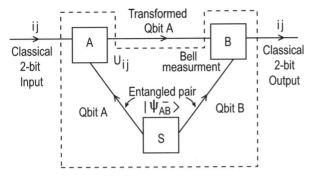

Fig. 2.6. Sketch of a scheme for quantum dense coding (see text for details). Two classical bits can be transported from A to B via the quantum information in a Qbit, equivalent to only *one* bit of extractable classical information. Any attempt by a "spy" to look into the messenger Qbit would destroy its quantum state and the related quantum information. S is the source of two maximally entangled Qbits in an asymmetric Bell state

2.9 Dense Coding, Teleportation and Quantum Information

Entanglement plays an important role in quantum computing. It also serves to point out some additional intriguing, counterintuitive properties of quantum systems. One example is *quantum dense coding* (e.g., [61]), a scheme in which one can use just *one* Qbit, from which normally one can extract only one classical bit, to carry *two* classical bits from one place to another, thus potentially doubling an information channel's transmission capacity. Figure 2.6 shows the arrangement. Alice wants to send to Bob two classical bits encoded in a pair i, j of binary digits, but for some reason (the need for secrecy for instance) she cannot do so classically. First, both agree on a way to label the four unitary operators shown in (2.13) and (2.21) with two binary digits. They choose, for instance

$$U_{00} = I \quad U_{01} = Z \quad U_{10} = X \quad U_{11} = Y. \tag{2.22}$$

Then Alice and Bob receive one Qbit each from an entangled pair prepared in the asymmetric Bell state $|\psi^-\rangle$. Alice performs a unitary transformation on her Qbit A with the operator that corresponds to the pair of digits to be transmitted, and sends her Qbit classically, i.e., as if it was a "gadget," to Bob (anyone trying to peek into it will destroy its superposed state!). The unitary transformation performed by Alice will have changed the *complete* state of the pair into one of the four Bell states according to relations (2.21). When Bob receives Alice's Qbit, he performs a Bell measurement on both particles to determine which of the four possibilities the complete state has assumed. The result will tell him what Alice's transformation was, i.e., what the two

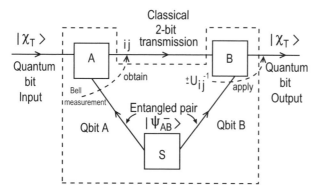

Fig. 2.7. Scheme of teleportation. A Qbit T in an unknown superposed state $|\chi\rangle_T$ at A can be regenerated at a distant point B by sending just two classical bits from A to B. S is the source of two maximally entangled Qbits in an asymmetric Bell state. The no-cloning theorem requires that the state of $|\chi\rangle_T$ at A be destroyed in the process – which it is

indices of the operator used have been. In this way, thanks to entanglement, a two-bit information was "compressed" reversibly into a device (Qbit A after the transformation) which, when independent, can really carry only one bit of classical information.

Next, consider the case of *quantum teleportation* (e.g., [61]). Call A and B two entangled Qbits, e.g., two electrons or photons emitted from source S in different directions (Fig. 2.7), and assume again that the pair is in a Bell state $|\psi^-\rangle$, (2.20a). Qbit A is received by Alice, Qbit B goes to Bob; both observers can be far away from each other. A *third* Qbit T (same type of particle as A and B) in a general superposed state $|\chi\rangle_T = a_T|0\rangle_T + b_T|1\rangle_T$ is available to Alice, but its quantum state is unknown to her (any measurement to find out would destroy it; the coefficients a_T and b_T could be determined by Alice only statistically with many identically prepared samples of T). The task of teleportation is to transfer Qbit T to Bob *without physically taking it there*. Stated in less science-fiction terms, the task is to enable Bob to create at his place a Qbit *identical* to T, but without him learning anything about its state. The no-cloning theorem (see previous section) tells us that this procedure necessarily will require 1. that the quantum state of T not be revealed at any time to either observer, and 2. that the Qbit in Alice's hand be destroyed in the process (otherwise we would have made an exact copy of T, which is prohibited).

To solve the problem of teleportation, we consider the complete state vector of the *three* particles:

$$|\chi\rangle_{ABT} = |\chi\rangle_{AB}|\chi\rangle_T = \frac{1}{\sqrt{2}}\left[|0\rangle_A|1\rangle_B - |1\rangle_A|0\rangle_B\right]\left[a_T|0\rangle_T + b_T|1\rangle_T\right].$$

With some algebra and taking into account the definitions (2.20) one can prove the following equality:

$$|\chi\rangle_{ABT} = |\chi\rangle_{AB}|\chi\rangle_T$$
$$= \frac{1}{2} \{ |\psi^-\rangle_{TA} [-a_T|0\rangle_B - b_T|1\rangle_B]$$
$$+ |\psi^+\rangle_{TA} [-a_T|0\rangle_B + b_T|1\rangle_B]$$
$$+ |\varPhi^-\rangle_{TA} [+b_T|0\rangle_B + a_T|1\rangle_B]$$
$$+ |\varPhi^+\rangle_{TA} [-b_T|0\rangle_B + a_T|1\rangle_B] \} .$$

The state of the system of three Qbits has now been made explicit in terms of the Bell states of the two particles T and A that are in the hands of Alice. Therefore, if Alice performs a Bell state measurement on her two particles, the complete state of the system will collapse into whatever basis state was revealed in her measurement, i.e., into one of the four summands above. The probability of obtaining each basis Bell state in the measurement is exactly $1/4$; thus this process provides her with no information at all about the Qbit T. The particle in the hands of Bob will jump into a state connected to whatever Bell basis Alice has obtained in her measurement in the above expression; indeed, one could view the brackets in the above expression as components of the complete state vector $|\chi\rangle_{ABT}$ in the Bell frame of the pair TA. The notable fact is that the new state of Bob's particle is related to the original state of Qbit T by one of the fundamental unitary operations U_{ij}, as one can verify by using (2.21) with the notations of (2.22):

$$[-a_T|0\rangle_B - b_T|1\rangle_B] = -U_{00}|\chi\rangle_T$$
$$[-a_T|0\rangle_B + b_T|1\rangle_B] = -U_{01}|\chi\rangle_T$$
$$[+b_T|0\rangle_B + a_T|1\rangle_B] = +U_{10}|\chi\rangle_T$$
$$[-b_T|0\rangle_B + a_T|1\rangle_B] = +U_{11}|\chi\rangle_T .$$

So, if Alice informs Bob (by classical means) which of the four Bell states she has obtained in her measurement (only *two* bits, the two subindices i, j are necessary to identify one of the four Bell states), all Bob has to do is perform the *inverse* unitary operation $\pm U_{ij}^{-1}$ on his Qbit, and presto! it will have been transformed into Qbit T. For instance, if Alice informs Bob that she found her two particles in the joint "01" state ($|\varPhi^+\rangle$), Bob applies $-U_{01}^{-1}$ to his Qbit and obtains

$$-U_{01}^{-1}[-a_T|0\rangle_B + b_T|1\rangle_B] = U_{01}^{-1}U_{01}|\chi\rangle_T = |\chi\rangle_T .$$

Anyone listening in on the open communication between Alice and Bob would not know what to do with the two bits obtained – they would have no subjective value to a spy! All this is not just the result of algebraic wizardry: It is a reflection of how quantum systems behave! No laws, neither classical nor quantum, have been violated (the communication between Alice and Bob was purely classical, and the original Qbit T was dutifully destroyed in the process). And yet there are monstrous counterintuitive implications, namely: With just two classical bits over a macroscopic channel, Alice was able to transfer to a distant site the infinite amount of quantum information contained in *two real numbers* (if we think of the state of Qbit T as expressed in the form (2.15)). How is this possible? By what means, through what channels, did those two real numbers reach Bob's particle? Again, the key point is that the information we are considering here is *quantum information,* which, as explained in the previous section, propagates with infinite speed and is intrinsically inaccessible, hidden in the system. We must understand the introduction of the Qbit $|\chi\rangle_T$ by Alice in terms of nonlocality. The moment it appears it "contaminates" the entire system instantaneously – but, again, not to worry: We are prevented from extracting the quantum information it contains (at the most we can squeeze out one classical bit if we make a measurement). We are even prevented from *thinking* about the two real numbers involved as if they were classical information, just as we were not allowed to mentally picture a photon traveling along one given path *or* the other in a Mach–Zehnder interferometer. This is all part of a deal struck with Nature and the price we must pay if we want to describe and manipulate the quantum world quantitatively and purposefully under the only terms our brain and laboratory instruments can handle. Let us conclude by repeating an earlier statement: Quantum paradoxes are paradoxes because of our unavoidable urge to represent mentally what happens *inside* the quantum domain with images and paradigms borrowed from our sensory experience in a classical, macroscopic world. But nothing that *transcends* physically or informationally from a quantum system into the macroscopic, classical domain is paradoxical or in any way contrary to the laws of classical physics.

It is outside the scope of this book to describe the impressive experiments – mostly tabletop setups (e.g., [26]) – that have been and are being performed to demonstrate the quantum effects described here, as well as those that represent first steps in the implementation of technological systems for quantum computation. The interested reader is referred to the increasingly voluminous literature (e.g., [1, 7, 9, 10, 82]).

Returning to quantum computing, let me end this chapter with some quotes from Mermin's pedagogically superb article [76], reproduced here in reverse order:

> ... one can be a masterful practitioner of computer science without having the foggiest notion of what a transistor is, not to mention how it works ... If you want to know what [a quantum computer] is capable of doing in principle, there is no reason to get involved in the really difficult physics of the subject ... [But] to understand how to *build* a quantum computer, or to study what physical systems are promising candidates for realizing such a device, you must indeed have many years of experience in quantum mechanics and its applications under your belt.

3 Classical, Quantum and Information-Driven Interactions

As we look around us we realize that much of the macroscopic part of the Universe with which we interact appears to be made up of discrete, localized "clumps" of matter – complex bodies which in general have clearly defined two-dimensional boundaries. Our senses of vision and touch are geared toward the perception of edges, patterns, texture and shapes of the boundary surfaces of every-day objects (let us use the term "form" to designate collectively these geometric/topological properties of objects), and our brain is endowed with the cognitive machinery to integrate (with the occasional help of the sense of touch, especially at the beginning of our lives) the perceived shape of a 2D boundary into a 3D image of the whole object to which it belongs, identify its position and orientation in space, appraise motion or change of form, estimate its size and inertia and, ultimately, find out its interactions with other bodies and ourselves and thus determine its meaning and purpose. Of course, we also perceive continuous things like the blue sky or a blowing wind, but it is mainly from features that mark a *change* that information is extracted by our senses. For the sense of vision, we are talking about changes in the spatial domain such as sharp one- and two-dimensional gradients in luminosity, color and texture leading to the concepts of form and pattern and to the awareness of 3D space, as well as the changes in time of such features, leading to the concepts of motion and to the awareness of the flow of time. In addition to "objects in space" there are also "objects in time": the discrete trains of acoustical waves to which our sense of hearing responds. Although we do hear continuous sounds, the most relevant acoustical information is extracted from temporal changes in the acoustical wave field. Our sense of hearing is able to determine minute changes in the sound field in such a sophisticated way that this became the most fundamental communication venue of human beings: speech.

In visual and acoustic perception, the animal brain uses any available information that helps expedite the extraordinarily complex process of recognition. Symmetries, periodicities, regularities are used to interpolate missing information or predict in the short-term what is about to come to facilitate the identification process. In other words, brains are geared toward exploiting the fact that many environmental features in space and time have lower algorithmic information content (see Sect. 1.4) than the Shannon information needed to describe them point by point and instant by instant, respectively. That is why the recognition and appreciation of regularity and organization is a natural element in brain operation. Since we humans can introspect our own brain function (Chap. 6), we view self-organizing systems – whose form and structure makes them more "predictable" – somewhat akin to ourselves, even when they pertain to the physical, inanimate domain. We marvel about a growing crystal and the singing tone of a whistling wind – yet these are all natural, logical and mathematical results of physical laws. Some systems are more structurally and/or functionally organized than others, to the point that certain very tiny regions of the Universe like Planet Earth have allowed for the development of highly self-organizing objects like viruses, bacteria, plants, animals and humans. As we shall discuss in this chapter and, more specifically, in Chap. 4, we human beings are at the end of the line (as we know it) of a natural process of global evolution, not just of the species, but of the whole Universe, that started with the Big Bang.

Brains of higher animals, endowed with memory for sensory events, are able to determine correlations between observed changes and build up a repertoire of subjectively experienced (but not always statistically verified!) cause-and-effect relationships between objects and events in their environment. The stored information on these relationships, which represents the acquired "knowledge," enables the animal to take advantageous behavioral action when confronted with external events in real time. Human brains can in addition mentally evaluate and order in time, or separate into specific categories, previously experienced cause-and-effect relationships, and seek out and explore new ones, even if current environmental and somatic conditions do not require this. Based on such internally generated or reordered information, human brains can make *long-term* predictions and plan appropriate behavioral response to environmental events long before these actually happen. Mental actions, both in animals and humans, involve at least two kinds of *correspondences*. One kind consists of correspondences of perceived forms, patterns, configurations and events in the environment with specific patterns of short-term neural activity and longer-term changes in synaptic architecture (representing the brain's input information processing). The other kind consists of correspondences between specific changes in real-time neural activity and the motor response of the organism (the brain's output processing). Human brains can, in addition, establish correlations among distinct classes of neural activity without any concurrent input from body and environment (the human thinking process). All higher animals are able to pair certain environmental constellations (a source of food, the sight or sound of a potential partner, a looming danger) with specific gestures, facial expressions and acoustical signals (again, a correspondence!) and thus communicate at a distance with other animals. Human beings have developed unique propositional language capabilities and are able to communicate to each other detailed facts about their own brain activity even if such activity is unrelated to current demands of organism and environment.

Why am I mentioning at this early time topics that are scheduled for detailed discussion only in the last chapter of this book? First of all, because I want to voice a deep but unavoidable frustration: In trying to write about the Universe I am unable to fully suppress the thought that what is doing the thinking and commanding the writing – my brain – is a nontrivial integral part of this Universe. Second, from a less subjective point of view and directly related to the concept of information, I mention these topics because I want to emphasize the importance of key concepts like form, pattern, order, symmetry, periodicity, change, correspondence, interaction, purpose and meaning, and their inextricable relation to the way how all living beings function, and how our brains extract and process information and construct mental images of the Universe. Third, we already have mentioned human brain function anyway, in Chap. 1 and Chap. 2. Finally, I want to call attention to the familiar but not at all trivial fact that patterns, forms and shapes crucial to the development and sustenance of life require the existence of one- and two-dimensional aspects of matter, such as polymers and giant carbon-based molecular strands at the microscopic level, and the clearly delineated surfaces of macroscopic bodies, respectively.

3.1 The Genesis of Complexity and Organization

Much of the known matter of the Universe is arranged into discrete regions separated by boundary layers whose thickness is small compared to their 2D extension. Aside from the normal objects in our immediate environment, this includes the tectonic layers in the interior of the Earth; the rather distinct layers of our atmosphere and magnetosphere; the highly variable plasma regions of the solar wind; the distinct regions of our Sun and other stars; the halos of mysterious dark matter of the galaxies – even the galaxies themselves are found to be distributed in the Universe along irregular filaments or "surfaces" which enclose huge bubbles of relative void. The late Nobel Laureate Hannes Alfvén, "father" of plasma physics, called it "the cellular Universe."

As the Universe evolved and expanded, the gradual condensation and stratification of materials after its explosive birth (the "Big Bang") led to increasing inhomogeneities and to an increasing complexity of their structure and the interfaces separating them. Although our book does not deal with cosmology, we must examine some aspects of physical and chemical evolution that are relevant to achieving a better understanding of the concept of information and the emergence of its role in Nature. They are: fluctuations and change; formation of "islands" of stability; the microscopic physical/chemical properties and macroscopic topology of the surfaces of condensed matter; and self-organization and emerging complexity.

We begin with the concept of change. The seeds of all change are fluctuations – not just on the grand cosmological scale, but on any scale from the Planck length up.[†] Every single elementary particle, nucleus, atom, molecule, star and planet is the deterministic result of some random event in the history of the Universe. Every transition from steady state to rapid change is triggered by fluctuations. This applies to a pot of hot water about to boil, to the emission of an alpha particle by a radioactive nucleus, to a "free will" decision by a human brain, or to a mass riot – perhaps even to the birth of the Universe itself. We already mentioned in Sect. 2.1 that at the quantum level fluctuations of any kind are allowed in empty space as long as they are compatible with Heisenberg's uncertainty principle (relation (2.1b)), i.e., as long as they are of short enough duration. In absence of free energy, they would remain unobservable. But immediately after the Big Bang, an enormous concentration of radiative energy was available and any uniform and symmetric distribution of matter and energy could be broken by such quantum fluctuations, ultimately leading to the formation of self-gravitating accumulations of visible matter such as stars, galaxies and clusters of galaxies. Indeed, the uneven distribution of matter and background radiation today may be a "map" of the distribution of random fluctuations that occurred during very early stages of the Universe – from a few nanoseconds to a few minutes after its explosive birth. This is somewhat related to what we stated in Sect. 1.4 regarding the constancy of algorithmic information (in the present case, it would be the number of bits needed to describe such a "map").

In general, we may distinguish two kinds of fluctuations. The first kind comprises transient events which decay with a timescale similar to that of their creation – they appear at random, decay and disappear. In a gas, individual collisions of molecules are random processes and whereas the actual evolution of the entire system at the microscopic level will fully depend on whether one given molecule went to the right or to the left in a particular collision, the macroscopic development of the gas will be left unaffected by that single fluctuation. There are fluctuations which individually leave no lasting trace, but collectively could produce observable effects under certain conditions. Examples are the creation and annihilation of virtual particle–antiparticle pairs which convert empty space into a dynamic medium at the quantum level; they cannot be observed individually but collectively they may influence the system in the presence of matter, such as conferring a magnetic moment to an electrically neutral particle like the neutron (see Sect. 3.4). At the mesoscopic level, we may point to the small transient changes of the pressure in a fluid due to local fluctuations in otherwise isotropic random

---

[†] The Planck length is defined as $l_P = \sqrt{(\hbar G/c^3)} = 1.616 \times 10^{-35}$ m, where G is the universal constant of gravitation (see Sect. 3.4). It is the smallest radius of curvature of a 4D Universe compatible with Heisenberg's uncertainty relation (2.1) and plays a fundamental role in cosmology and string theory.

molecular motion, giving rise to the jitter of small bodies suspended in the fluid (Brownian motion).

The second kind comprises fluctuations that surpass some critical size and, with the supply of some free energy, trigger instabilities which grow and lead to macroscopic consequences. Some of these are determined by physical laws alone; in others randomness participates to some degree in form and structure (e.g., see [88]). Familiar examples in our macroscopic environment are the growth of a crystal in a locally perturbed supersaturated solution, the metastable convection cells (Bénard cells) in an anisotropically heated fluid (e.g., the rising air under cumulus clouds), the vortices in a turbulent flow, etc. In the two last cases, fluctuations in the random motion of molecules become amplified and organize the fluid into cells of macroscopic flow. As noted above, during the early expansion of the Universe, fluctuations were responsible for the breakdown of uniformity and symmetry and the formation of compact and stable structures of matter of increasing complexity and size – from elementary particles to nuclei, atoms and molecules, to condensed matter, stars and planets. The history of the evolution of the Universe since the Big Bang is one of sporadic appearance of localized random fluctuations followed by a deterministic organization of matter by the rule of natural law. We should examine in more detail the intervening processes.

An underlying concept is that of the *selective advantage* of certain structures vis-à-vis others – configurations we shall call "attractors" (borrowing the term from chaos theory). Recall that we mentioned the great fragility of any artificial quantum system prepared in the laboratory such as a Qbit (Sect. 2.7): Decoherence caused by unavoidable interactions with the environment (fluctuations!) will ultimately break up a superposed or entangled state of such an artificial quantum system (see Sect. 5.6). There are, however, states that are robust and immune to tenuous external influences. If we imagine a gradually expanding and cooling "primordial cloud" of electrons and protons, random fluctuations in the collisional configurations would occasionally bring proton–electron pairs into a bound, stable configuration: the eigenstates of the hydrogen atom (e.g., [11]). Any other imaginable configuration would eventually decay – we are, indeed, in presence of a "natural selection" process that is random but has deterministic results. We can envisage a similar process in an expanding "cloud" of unconfined quarks (which happened earlier in the history of the Universe than the expanding cloud of protons and electrons; see below): Random fluctuations would eventually bring triplets of so-called up- and down-quarks together into a confined state forming stable nucleons (protons and neutrons); other configurations would decay with a lifetime shorter than the collision time. Thus, in the primordial subatomic world stable configurations emerged as the "survivors" in a sea of chaotic fluctuations: first the elementary particles, then atoms and later molecules. Even the fact that most antimatter has disappeared early on during the evolution of the Universe is attributed to a fluctuation in the total number of quarks and antiquarks which left a tiny superabundance of a factor of only 10^{-9} of matter over antimatter; the small excess survived, the huge rest of quark–antiquark pairs annihilated at a very early stage of the evolution (see for instance [101]), giving an additional boost to radiative energy.

Responsible for the very limited classes of stable configurations are elementary *physical interactions:* the so-called strong interactions between quarks to form nucleons; short-range strong and weak interactions between nucleons to form nuclei; electromagnetic interactions between nuclei and electrons to form atoms; electromagnetic interactions (electron exchange forces) between atoms to form molecules; and gravitational interactions between all matter. The forces binding the elementary particles into these "attractor" structures are governed by basic physical laws; their relative strengths are determined by universal constants of interaction (the coupling constants). The values of the latter, together with the masses and charges of the elementary particles,[†] are to be blamed for all aspects of our Universe – including our own very existence – except for the spatial distribution of matter which may be linked to the randomness of the primordial fluctuations. The "clock" which is timing the succession of all these processes is the expansion of the Universe, which led to a cooling and average density decrease, allowing the scale of fluctuations to slide over critical windows favorable to the formation of more stable and more complex structures.

Agglomerations of atoms and molecules gave rise to condensed matter and macroscopic bodies, under the control of *gravitational interactions.* To understand the role of fluctuations at this much larger spatial scale, imagine a cloud of neutral atoms with a uniform distribution filling all available space, in absence of external influences. If somehow there is a random temporary increase in the local particle number density, a small regional anisotropy will develop in the gravitational field. If some threshold in spatial and temporal dimension is surpassed, the particle distribution will begin to self-gravitate and attract more atoms from the surroundings to that area: The cloud will condense toward where the original fluctuation took place. If several such fluctuations occur throughout the cloud, several condensed regions will form. Fluctuations in the velocity distribution of the original atoms in turn will lead to collective motions, including rotation, of these regions. Galaxies and stars are thus formed as "attractors" (now in the literal sense!) in the gravitational field: metastable distributions of compact matter, originally triggered by some primordial fluctuations. Increasingly realistic numerical simulations are being conducted showing how such a process indeed generates galaxies with a filamentary or membrane-like distribution in space. On a much smaller scale and much later in the evolution of the Universe, fluctuations in rotating plasmas, where the magnetic field is now the organizing agent, lead to

---

[†] Which one day may be *deduced* from a few more fundamental constants in a "Grand Unified Theory."

macroscopic reconfigurations of currents – the so-called dynamo processes responsible for the self-sustaining internal magnetic fields of stars (like our own Sun) and metallic-core planets (Earth, Mercury and the Giant Planets) and moons (Jupiter's Ganymede).

In the macroscopic domain the two fundamental interactions that "survive" from the microscopic scale because of their long range are the gravitational and electromagnetic forces (Sect. 3.2). Under both gravitation and the interatomic or intermolecular forces (which are electromagnetic, too, but in their collective action are of shorter range), matter condenses and solid bodies are formed, mineralize, stratify – and environments appear containing limited-sized objects with distinct surfaces, regularities, symmetries and other forms of higher organization. Thanks to peculiar physicochemical properties of the molecules involved, one particular class of carbon compounds takes off on a development of its own, not just recently on Earth but long before on the surface of icy comets and planetary moons. Indeed, there is an ubiquity of simple organic molecules such as HCN, light hydrocarbons, even simple amino acids and polymers, which in the favorable environment of Earth have developed the capacity of information-driven interactions (see Sect. 3.5) and given rise to living organisms (Chap. 4) – membrane-encapsulated "islands" of self-organization with reproductive capacity, able to maintain a low-entropy state in metastable equilibrium with the environment. Darwinian evolution thus emerges as an integral, highly localized, but nonetheless "explosive" part of cosmic evolution and differentiation under certain very specific environmental conditions and availability of energy. Despite the enormous complexity of the relevant biomolecules, there is a rather limited repertoire of organic building blocks – it is the variety of their order and related *patterns* that makes up the diversity: Form and pattern begin to displace energy in the driving of evolution (Sect. 3.5 and, e.g., [52]). Still, there is a big vacuum in the understanding of the transition between simple organic molecules and large bioactive compounds, and the likely catalytic action of the surfaces of certain minerals in an aquatic environment. We shall come back to this in Chap. 4.

In all this we must be careful not to forget that while the occurrence of fluctuations with long-term consequences is stochastic, once an effect on their surroundings has been "triggered," determinism takes over, guided by the laws governing the intervening physical interactions – selective advantage determines the course of evolutionary change. Table 3.1 schematically shows the classes of "attractors" – the Universe's stable or quasistable building blocks and prototype structures – that emerged during expansion and cooling as the result of random fluctuations at various scales. I want to remind the reader that cosmology is not within the purview of this book, so let me just comment on some of the "milestones" of the history of the Universe (e.g., [24, 101]) listed in Table 3.1. There are two basic periods: One was the *radiation-dominated* epoch spanning from the first Planck time of 10^{-43} s (the

Table 3.1. Stable "attractor" structures emerging as the Universe expanded and cooled, as the result of random fluctuations and the effects of deterministic physical laws governing basic physical interaction mechanisms. They represent increasingly complex and organized entities, if we define "complexity" by the number of constituents and "organization" by emergent stability and regularity in structure and dynamics. The evolution of life is shown as a continuation of the increasing organization and differentiation in the Earth's environment – a result of the emergence of information-driven interactions. For details see text, as well as Sect. 3.5 and Sect. 4.5

Ensembles	Force-field quanta or type of interaction	"Attractor" structures	Temperature ($1\,\mathrm{GeV} \approx 10^{13}$ K)	Time after Big Bang
Radiation-dominated, opaque to photon gas				
Initial chaos	Grand unified force		$> 10^7$ GeV	$< 10^{-24}$ s
Radiation	High-energy photons weak bosons	Particle–antiparticle pairs (leptons, unconfined quarks)	~ 100 GeV	$\sim 10^{-24}$ s
symmetry breaks, weak interactions begin most antimatter disappears dark matter forms (?)				$\sim 10^{-9}$ s
Quarks	Gluons	Nucleons, hyperons (confined quarks)	1 GeV	$\sim 10^{-4}$ s -1 s
Nucleons	Gluons, weak bosons	Light nuclei	1 MeV	$\sim 2 \times 10^2$ s
Light ion–electron plasmas	Photons	Atoms	eV	$\sim 3 \times 10^4$ a

time it takes a photon to travel a Planck length, see footnote † on page 82) to about 300 000 years, followed by the *matter-dominated* epoch until the present ($\sim 14 \times 10^9$ years after the Big Bang). The initially expanding fireball was "pure radiation" of ultra-high energy ($> 10^7$ GeV). At these energies photon–photon collisions led to the creation of particle–antiparticle pairs (see Sect. 3.4), and after about 10^{-24} s emerged a nearly uniform and extremely hot gas of leptons (electrons, muons, taus and neutrinos), quarks (the fundamental constituents of hadrons like protons, neutrons, kaons and hyperons) and what is called gauge bosons (particles like photons, gluons, $W^\pm$, Z_0, responsible for elementary particle interactions), rapidly expanding and cooling from the initial Planck domain. However, at these energies the coupling strengths were probably all equal (no distinction between strong, weak and electromagnetic interactions – maybe even gravity); the gas was only weakly interacting and quarks were "free" or unconfined (see Sect. 3.4). After about 10^{-9} s the temperature had decreased sufficiently, and electroweak interactions appeared, mediated by the massive $W^\pm$ and Z_0 particles. Once the temperature fell below about 1 GeV ($= 1.602 \times 10^{-10}$ J $\approx 10^{13}$ K) after about 10^{-4} s, the quark–gluon plasma entered a strongly interacting phase.

Table 3.1. continued

Ensembles	Force-field quanta or type of interaction	"Attractor" structures	Temperature ($1\,\mathrm{GeV} \approx 10^{13}\,\mathrm{K}$)	Time after Big Bang
Matter-dominated, transparent dark energy accelerates expansion				
Atoms, ions, electrons	Gravitons	Galaxies, stars, heavy nuclei		$\sim 4 \times 10^5$ a
Atoms	Electron exchange force	Molecules		
Atoms, Molecules	Gravity, chemical	Cold condensed matter		
Planetary differentiation				
Condensed matter	Chemical	Objects with 2D surfaces		
Self-organization, macroscopic complexity increases				
C-based compounds	Chemical	Biomolecules catalysis on solid surfaces		$\sim 10^{10}$ a
Information-based interactions appear and accelerate differentiation on Earth				time BP
Prebiotic molecules	Evolution	Viruses, cells		1.3×10^9 a
Cells	Darwinian evolution	Multicellular organisms		$\sim 800\,\mathrm{Ma}$
Multicellular organisms	Darwinian evolution	Plants, animals		$\sim 410\,\mathrm{Ma}$
Animals	Darwinian evolution	Humans		$\sim 6\,\mathrm{Ma}$

Quarks and antiquarks annihilated creating an additional burst of radiative energy. Because of the small initial relative superabundance ($\delta N/N \approx 10^{-9}$) of quarks over antiquarks, the remaining quarks bound into stable triplets to form baryons, including protons and neutrons. Between 1 s and 100 s, the temperature had fallen to a few MeV (10^{10} K), unstable mesons and hyperons disappeared, nucleosynthesis started and light nuclei appeared on the scene.

The end of the radiation-dominated epoch came after about 300 000 years, when fluctuations in the electron–ion plasma led to the formation of stable atoms; the temperature is now only about 10^4 K (a few electronvolt). Due to this condensation of matter the Universe became transparent to photons, which are now able to carry "low grade" energy and entropy away from the increasingly complex "clumps" of matter. The gravitational field began its crucial role with the formation of galaxies and stars about a million years after the Big Bang (but remember, the "seeds" for the formation of self-gravitating matter may already have been sown by quantum fluctuations when the Universe was only 10^{-35} s old!). Inside stars, nuclear reactions fused lighter nuclei into heavier ones – among them the crucial element carbon, formed by fusion of three helium nuclei (a highly unlikely process were it not for the existence of a strong resonance in the cross section at exactly the temperature existing in certain stellar cores). Our Solar System and its planets formed later, about 10^{10} years after the initial explosion. At some as yet unknown moment at the beginning of the stages shown in Table 3.1 *dark matter* and dark energy appeared.[†] Current high-energy particle experiments and observations of 1. the expansion rate of the Universe during the last 6 billion years, 2. the distortion (lensing) of images of very distant galaxies as their light passes through the periphery of galactic clusters, and 3. the rotation of stars in the outer galactic arms around the galactic center (which shows a nearly constant linear velocity instead of the Keplerian radial dependence that would correspond to the distribution of visible matter), all point to the existence of a vast reservoir of invisible matter and energy. Theory postulates as candidates for dark matter the so-called WIMPs (weakly interacting massive particles) and axions; none have been detected yet. The fractions of visible matter and detectable energy have now been demoted to only a few percent of the total mass and energy of the Universe.

The entities shown on the right side of Table 3.1 are increasingly complex and organized, if we loosely define complexity by the number of interacting constituents of a stable structure (see Sect. 1.4), and organization by the emergent properties of regularity (in structure and temporal evolution) – all as the result of the forces responsible for the respective systems' formation. Let us take a more general look at this increasing organization at the macroscopic level. There is no incompatibility with the second law of thermodynamics because these "islands" of increasing or sustained order are spatially confined yet thermodynamically *open* systems, to which free energy can be fed and/or from which "low grade" energy can be extracted. Let us remember that the traditional version of the second law of thermodynamics applies only to *closed* systems (see Sect. 5.4).

The formation and increase of local order and regularity in an open system arising through physical processes is called *self-organization* (Fig. 3.1); it requires a regular inflow of free energy from its environment and a disposal of entropy to the outside [24]. In this way the second law is upheld: Localized entropy decreases occur at the expense of a greater entropy increase in the rest of the Universe, so that the sum total always increases. It is important to distinguish an open system with systematically growing and/or sustained organization from a closed system that is found in a transient, low entropy state caused by a fluctuation in its microscopic structure. As we shall discuss in Sect. 5.3, such fluctuations are not prohibited by any physical law; for instance, it is not at all prohibited for the molecules of a gas to assemble in one corner of a closed container. But the probability for this to happen without external intervention (a piston, for instance) is extraordinarily small.

[†] The two most important areas of present-day research in cosmology are focused on the unification of all interactions and the role of string theory in the creation of matter (fermions and bosons) during the ultra-hot initial phase, and on the present-day nature of dark matter and dark energy.

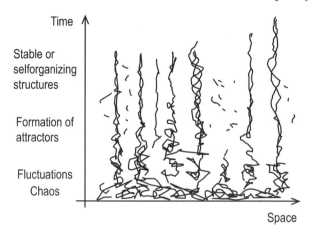

Fig. 3.1. Schematic view in "spacetime" of the formation of organized, stable entities triggered by fluctuations in the early Universe

And even if it were to happen, it would last only an extraordinarily short interval of time. In self-organization, a continuous supply of free energy from outside and the extraction of heat (low grade energy) are necessary to increase and maintain organization and a nonequilibrium state of low entropy (Sect. 5.3). Self-gravitation and the appearance of inhomogeneities (stars, galaxies) invariably involve dissipation of energy – mainly because of the attractive character of the gravitational field (see Sect. 3.2). Gravitational energy is converted in such a way as to increase the number of degrees of freedom of the entire system (heat dissipation, emission of low-energy photons, or formation of heavy nuclei); as a result, a uniform distribution of gravitating matter will have a *lower* entropy than a clumped distribution – contrary to what happens to a distribution of nongravitating gas molecules (see Sect. 5.5). The absolute "end of the line" of self-gravitating matter is compression into a black hole (the most radical nonuniform distribution!), which represents a state of maximum entropy [5]; the "end of the line" of the molecules in an ideal gas, instead, is a constant density distribution (the most radical uniform distribution!), which represents the state of maximum entropy in this case (Sect. 5.5). An important role is played by a slowly varying (quasistatistic) magnetic field in the self-organization of cosmic plasmas, which tend to assume a configuration of lowest entropy – at the expense of dissipation of energy in form of radiation or collisional interactions with a neutral gas.

According to our definition of the degree of complexity in Sect. 1.4 and its relation to algorithmic information, we may restate what we have said earlier concerning the history of the Universe: We witness a generation of algorithmic information in random events, followed by preservation of this information during phases of self-organization ruled by natural law. Indeed, we may be tempted to state with *Chaisson* [24] that "the process of cosmic evolution is continuously generating information." But we must then ask right away: Are the physical, nonbiotic evolutionary processes and respective interactions really *using* this information? Did this continuously generated information have a purpose and meaning at the time it was generated, or does it have purpose and meaning *only now* for an intelligent being who is observing and *studying* the Universe [97]? We raised this question in the Introduction and will deal with it again in Sect. 3.3 and Sect. 3.4, and in Chap. 5.

We mentioned above the selective advantage of some particular component structures in terms of their stability or "survivability" in an evolving Universe. Other structures are possible and may have formed during the course of evolution, but if they did, they were evanescent and did not lead to any stable "islands" of increasing organization (like the above mentioned gas that assembles in one corner of the vessel). Today, however, we can produce new isotopes, new organic molecules, new nanostructures, new chemicals, new breeds and clones, even new virus-like or cell-like entities that have *not* undergone any natural, selective evolution (and we can willfully compress the gas into a corner with a piston). These actions have all been planned and designed with a premeditated goal – a very different process, requiring the intervention of a human brain. Here, indeed, we must appeal to the concept of *information* and its intervention in the physical world! Some of these products may on occasion arise naturally: Transuranian isotopes surely are being formed in the interior of massive stars, but they decay; C_{60} fullerene molecules may appear in volcanic ash, but they would quickly disintegrate; certain giant "bio"-molecules manufactured today for industrial purposes may have appeared on occasion in natural processes, but they would not have been metabolized or been used in any way by a natural biological organism. It is human intention and action that is now generating and preserving these new products – structures that emerge not from physical laws alone but from brain information processing which causes deliberate, planned changes in the initial conditions that enter in the laws governing the systems under consideration (see details in Sect. 5.2 and Chap. 6).

3.2 Classical Interaction Mechanisms

When two bodies approach each other they may *interact,* that is, change their motion, shape, constitution or any other property. This happens with elementary particles in the quantum domain as well as with macroscopic bodies. Interactions between particles are what under the right circumstances or *initial conditions* may lead to stable, bound structures, like the ones listed on the right side of Table 3.1. If the initial circumstances are *not* right, on the other hand, the interaction may lead to no new structures, or to unstable ones which after a time will decay. Quantum fluctuations of the vacuum and/or fluctuations in particle number density or velocity enable the chance appearance of favorable initial conditions. In this section we examine the different types of interactions responsible for the creation of new "next generation" structures with emergent properties that represent more than the sum of the individual properties of their components.

I will not attempt a formal definition of interaction process; rather, we shall take it to be what is called an "epistemological primitive" and use the "working definition" given at the beginning of the preceding paragraph. We are mainly interested in the fundamental characteristics of the process, and for that purpose will begin by reviewing in a rather pedantic detail some very elementary and old physics. This seems ridiculous, taking into account that in Chap. 2 we have already dealt with equations describing quantum systems. But we should remember that the entire framework of physics, indeed, the dynamics of the whole Universe, is based on the properties of a few types of interactions between two particles, the linear superposition of their effects, and their macroscopic expression in the world around us. This surely warrants a closer inspection of the elementary concept of interaction between two bodies. In what follows we shall mainly focus on those aspects that later will be crucial to our quest for a more objective definition and a better understanding of the concept of information.

As is done with most physical theories, to lay down the basic laws we first formulate an idealized model of the system under consideration (Sect. 5.2). In our case it consists of two isolated "material points" in the classical non-relativistic realm (macroscopic bodies of a size negligible compared to their mutual distance; rotational energy negligible with respect to translational energy; velocities negligible with respect to that of light), isolated from the rest of the Universe (i.e., under assumed negligible external influences). We further assume that a rigid Euclidean reference system exists (with negligible effects on our material points) with respect to which we measure position, and that at each point of space we have at our disposal a clock at rest with respect to the reference frame and synchronized with all other clocks. We also assume that we can measure the position of our material points and the local time at which the measurement takes place without in any way disturbing our (classical) system. This is a very tall order of conditions; I point them out in such detail, because they will be examined again in Chap. 5 in a discussion of how the human brain intervenes in the preparation and description of a physical system and how this fact (but not Nature itself) brings the concept of information into physics.

Next we conduct a series of "ideal" experiments involving "ideal" purely kinematic measurements with our "ideal" system of two material points. Ideal experiments must be possible "in principle" but are difficult to carry out in practice. The goal is to arrive at some relationships between physical quantities which can be verified with some more realistic measurements and with sufficient precision to certify what the ideal experiments claim. Under these guidelines, we prepare our system of two material points by placing them at

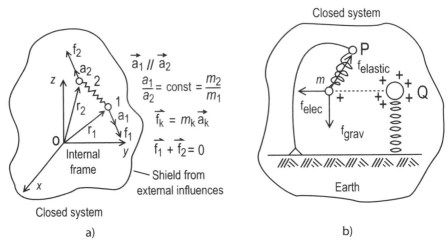

Fig. 3.2. (a) Two isolated material points in mutual interaction: fundamental relationships. (b) One material point simultaneously subjected to the linear superposition of three different interactions with a massive body (P and Q are "part of Planet Earth")

points r_1^0 and r_2^0 (r: radius vector pointing from the origin to the point in question) with initial velocities v_1^0 and v_2^0 (which could be zero) with respect to the chosen frame of reference (Fig. 3.2a). We observe that one of three things may happen. 1. The bodies continue with constant velocities in rectilinear motion, in which case we say that the material points are free of mutual interaction *and* that the frame of reference is an "inertial frame." 2. They suffer instantaneous accelerations with parallel acceleration vectors $\boldsymbol{a}_1$ and $\boldsymbol{a}_2$ pointing in mutually opposite directions, the ratio of their moduli a_1/a_2 being constant for the two material points in question and independent of mutual position and time, depending only on the two bodies in question. In this case we say that both bodies are in mutual interaction and that the reference frame is, again, inertial. 3. Same as 2, but without fulfilling the conditions on the direction and the moduli of the accelerations mentioned for 2. In this case we say that we are located in a "noninertial" frame of reference, or that the system of the two material points is not isolated (i.e., that it is subjected to outside interactions).

If we introduce two scalars m_1 and m_2 such that $a_1/a_2 = m_2/m_1$, we can express both experimental results 1 and 2 above in vector form, valid at all times and for any kind of mutual interaction:

$$m_1\boldsymbol{a}_1 + m_2\boldsymbol{a}_2 = 0. \tag{3.1}$$

Each term is called the *force* $\boldsymbol{f} = m\boldsymbol{a}$ on the corresponding material point due to the mutual interaction. The scalar m pertaining to each material point is called *inertial mass;* under the nonrelativistic conditions imposed it is a constant that only depends on the body in question. Thus defined, inertial mass is an independent physical magnitude; the unit of mass must be defined physically by choosing one particular, internationally accepted, body as the standard. Relation (3.1) leads to the universal "action–reaction" relation $\boldsymbol{f}_1 + \boldsymbol{f}_2 = 0$ as a mere consequence of the definition of force introduced as a dependent physical magnitude, valid for any kind of interaction in an inertial frame of reference (Fig. 3.2a). Further ideal experiments demonstrate that when several different interaction mechanisms act simultaneously on one of the bodies, or when there are more than two material points in interaction, the total force on one of the mass points is the vector sum of the individual interaction forces; in other words, interactions act independently from each other and their effects add linearly (Fig. 3.2b).

This in essence represents Ernst Mach's original reformulation of Newton's laws (see [70], p. 304). It has the advantage that it does not involve a circular definition of force and mass; note that both magnitudes are determined on the basis of purely kinematic (geometric) measurements. Relation (3.1) is quite difficult to verify with precision in the laboratory under the conditions described, but its consequences, such as the conservation of total linear momentum (integration of (3.1)),

$$m_1 \boldsymbol{v}_1 + m_2 \boldsymbol{v}_2 = m_1 \boldsymbol{v}_1^0 + m_2 \boldsymbol{v}_2^0 = \text{const} \qquad (3.2)$$

can be tested easily, for instance, with elastic collisions between mass points (this is indeed how the masses of many elementary particles have been determined). The conservation of angular momentum also follows directly from (3.1) (and leads to Kepler's long-verified second law of planetary motion).

Relation (3.1) applies to two interacting material points in "isolation from the rest." This means that they are not interacting with any other body. Since this is physically impossible because we cannot shield the system from gravitational interactions with the rest of the Universe (see next section), we have to define "isolation" as "negligible interaction with other bodies," so that the respective forces can be neglected vis-a-vis the force ma arising from the two-body interaction. To maintain one of the interacting bodies at rest, we must put it in interaction with a massive *third* body (or set of bodies) in such a way that all forces acting on it are in equilibrium (as in Fig. 3.2b). The entire framework of classical Hamiltonian mechanics can be derived from Mach's formulation ((3.1) and the vector sum of interaction forces). It applies to a system of interacting mass points, subjected to given constraints that impose preset geometric limitations on their motion (with the physical mechanisms of constraint not considered as part of the system). Given a system of N material points at an initial time t^0 located at coordinates q_k^0 (which need not be Cartesian but must be compatible with the constraints) with associated momenta $p_k^0 = m_k \, dq_k / dt|^0$, Hamiltonian mechanics allows the determination of the system's coordinates and momenta at any other time t. In other words, it establishes a *correspondence* between an *initial state* of the system $\{q_k^0, p_k^0\}$ at time t^0 and a *final state* $\{q_k, p_k\}$ at time t. It is important to note that in absence of frictional constraints, the system is perfectly time-reversible (because the fundamental equations are): If we run the system "backwards" taking as initial state what was the final state but with reversed velocities or momenta $\{q_k, -p_k\}$, it will arrive after a time t in the state $\{q_k^0, -p_k^0\}^\dagger$ (this is the same as running a movie of the system backwards). An apparently trivial statement, to which we will return later (Sect 5.2), is the following: For a given system of mass points it is *our choice* when the initial conditions are set to occur, and what these are to be (either theoretically or experimentally). The time t marking the final state is our choice, too – but it is *not* our choice what the conditions of the system will be then. You will not say: How come, are not the positions and velocities of, say, the planets at any chosen initial time t^0 really given by Nature and not by us? This is a crucial point regarding the role of information in physics, to be discussed in Sect. 5.2.

We now turn to the most important classes of interactions in the classical, macroscopic, nonrelativistic domain. Gravitational interactions act between all material bodies; electromagnetic interactions act only between certain types of elementary particles or between macroscopic bodies that have been prepared in a certain way (electrically charged or magnetized). There are also elastic interactions, requiring the action of a medium or a special physical device between the interacting bodies (of negligible mass or considered to be an integral part of one of the interacting bodies). In the macroscopic domain there are "collective" interactions like thermal, frictional and chemical which, however, are ultimately based on electromagnetic interactions at the atomic or molecular level.

Each class of interactions must be quantified through measurements in ideal experiments. For that purpose, we place one of the bodies, designated a *source,* at a fixed point such as the origin of the inertial frame of reference – which means that we must assume that it is rigidly attached to an inertial mass much larger than that of the other body with which it interacts, so that according to (3.1) its own acceleration is always $a \to 0$. We then use the other body as a *probe* to explore the dependence of the strength of the interaction as a function of space. One verifies that the instantaneous force on the probe depends on some particular properties of the probe and on its instantaneous position $\boldsymbol{r}$.‡ The concept of *field* is then introduced as a *local property of space*

---

† Note that we said "time-reversible," yet we have *not* reversed the time – we reversed the velocities (momenta) and ran the time forward during a lapse $t - t_0$! We shall clarify this remark in Sect. 5.2.

‡ In all interactions except the magnetic one between moving particles (see further below) the measurements on the probe can be made under static conditions by equilibrating the interaction force with a force of different, but known, nature. In magnetostatics one can conduct *static* experiments at the macroscopic level with constant, closed currents and current densities (and with magnets, as it was done first historically).

surrounding the source, responsible for the force that acts on a particle of similar type when it is placed in that location of the field. Like with Mach's relations (3.1) one conducts a series of ideal experiments (difficult to do, but whose consequences are amenable to precise measurement) to determine how the interaction force depends on the probe, its spatial position, and on the properties of the source.

For the *gravitational interaction* one finds that, given two small probes 1 and 2 and placing them successively at given points of the space surrounding the source body at O, the respective interaction forces $\boldsymbol{f}_1$ and $\boldsymbol{f}_2$ at each point have always the same direction (towards O) and their magnitudes have a constant ratio which does not depend on the point in question but only on the two probes: $f_1/f_2 = \text{const}$. Introducing a positive scalar μ called *gravitational mass*, one can write $f_1/f_2 = \mu_1/\mu_2$ so that the vector $\boldsymbol{g}(\boldsymbol{r}) = \boldsymbol{f}/\mu$ is independent of the body used as a probe; it is called the *gravitational field* at the given point $\boldsymbol{r}$. We interpret the field as a property of the *space* surrounding the source, independent of the body (the probe) used to explore it. As in the case of the inertial mass, it is necessary to adopt a standard body as the unit of gravitational mass. For a point-like source body one further finds that $\boldsymbol{g} = -K_{\mathrm{s}}\boldsymbol{r}/r^3$ where K_{s} is a positive scalar whose value depends only on the source body. Assuming for a moment that the source does not have an infinite inertial mass, we can interchange the source with the probe, and the action–reaction relation, which is valid for any type of interactions under the present circumstances, will tell us that K_{s} must be proportional to the source's gravitational mass μ_{s}. Thus $\boldsymbol{g} = -G\mu_{\mathrm{s}}\boldsymbol{r}/r^3$ where $G = 6.61 \times 10^{-11}\ \mathrm{m^3 \cdot kg^{-1} \cdot s^{-2}}$ is the universal constant of gravitational interaction, a quantity that does not depend on anything else; it represents the strength of the interaction.

Instead of working with the gravitational vector field, one can introduce the gravitational *potential*, a scalar field $V(r) = -G\mu_{\mathrm{s}}/r + C$, from which $\boldsymbol{g}$ can be derived as a gradient: $\boldsymbol{g} = -\boldsymbol{\nabla}V$. The arbitrary constant C (value of V at infinity) will not influence the physical entity $\boldsymbol{g}$. The advantage of V is not only that it is a scalar (note the compression of information, Sect. 1.4!), but that $-\mu\Delta V = \mu(V_1 - V_2)$ represents the potential energy difference of a small body of gravitational mass μ between points 1 and 2. Combining this with Mach's relations (3.1) one obtains the theorem of conservation of mechanical energy, valid for any (curl-free) force field that derives from a scalar potential ("conservative" force fields):

$$V(r) + \frac{1}{2}mv^2 = E_{\mathrm{total}} = V(r_0) + \frac{1}{2}mv_0^2 = \text{const}. \tag{3.3}$$

In the expression of $V(r)$ the infinite value for $r \to 0$ in the expressions of the field and the potential of a point source does not bother in the classical domain: Our experimental conditions do not allow us to get "very close" to the source body – if we do, it would no longer appear as negligibly small! Experiment shows that gravitational masses and fields are linearly *additive;* so the gravitational field or potential of an extended body are the sum (the integral) of individual contributions from each infinitesimal element of mass of the body. Finally, a most fundamental *additional* experimental fact is that the gravitational mass of a body is proportional to its inertial mass so that the gravitational acceleration of a small body of inertial mass m is always proportional to the local $\boldsymbol{g}$.[†] This is why one does not distinguish between the two, setting $\mu \equiv m$ (which implies that we have chosen the same standard body to represent the units of both, inertial and gravitational mass). Note that gravitational forces are always attractive with the consequence that there is no way of shielding a given region from a gravitational field – a hint that the gravitational field is a property of "spacetime itself" (its curvature, e.g., [78]). This is also the reason why "completely isolated" systems do not exist.

For *electrostatic interactions* between charged bodies held at rest by some other types of interactions, one can conduct a similar set of ideal experiments and obtain functionally equivalent results. One is thus led to the electric charge q as an independent physical magnitude (which, contrary to the gravitational mass can be positive, zero or negative; the unit of charge and its sign are adopted by an international agreement) and to an electric field vector $\boldsymbol{E} = \boldsymbol{f}/q$ and the related scalar "Coulomb potential" $V(\boldsymbol{r}) = K_{\mathrm{e}} q_{\mathrm{s}}/r + C$, where q_{s} is the charge of the point source and $K_{\mathrm{e}} = (4\pi\varepsilon_0)^{-1} = 9.00 \times 10^9 \, \mathrm{kg \cdot m^3 \cdot s^{-2} \cdot C^{-2}}$[‡] is the universal constant of electrostatic interactions (ε_0 is the traditional vacuum permittivity). Like in the gravitational case, the electric field and charges are linearly additive. It is easy to charge macroscopic bodies electrically and measure their electrostatic interactions in the laboratory, but it is extraordinarily difficult to measure their mutual gravitational interactions. The reason is the great difference in the magnitudes of the respective constants of interaction: $G/K_{\mathrm{e}} = 7.42 \times 10^{-21} \, \mathrm{C^2 \cdot kg^{-2}}$ (which, by the way, means that the gravitational attraction between, say, two protons

---

[†] If this proportionality were not true, we would be living in a very different world: Different inertial masses would not have the same acceleration at a given point of the gravitational field, planets would move very differently (Kepler's third law would include the planets' mass ratio) and there would be no Einstein principle of equivalence. One way of stating this principle is to say that "it is impossible to determine with experiments carried out exclusively inside a closed system in which relation (3.1) has been verified, whether it is an inertial system or a system accelerated with acceleration $\boldsymbol{g}$ in a uniform gravitational field of intensity $\boldsymbol{g}$" – an astronaut born, raised and educated in an orbiting space station without windows would have no way of determining through indoors experiments if he is living in an inertial system far from any star, or if his system is accelerated, orbiting or falling in a gravitational field! (All this is strictly true only if the region in which the experiments are carried out, e.g., the space station, are infinitesimally small, so that "tidal" effects of gradients of the field can be neglected.) This is a dramatic example of how the numerical value of one single parameter in a physical law shapes the entire Universe (in a more general expression $\mu = C m^{\alpha}$ we have $\alpha = +1.000$, tested to 1 part in 10^{11}).

[‡] We shall use rationalized SI units throughout the book.

is $\approx 10^{-36}$ times that of their electrostatic repulsion). We are familiar with the gravitational interaction only because the mass of the Earth is so large!

For *elastic interactions,* there is no equivalent of "charge" because the elastic force does not depend on either of the two interacting bodies, only on the interaction mechanism. The "field" is represented by a central potential $V(r) = -\frac{1}{2}kr^2 + C$ in which k is the *elastic constant,* which depends on the particular interaction setup in question. Elastic interactions are important in condensed matter physics, acoustics and molecular physics;[†] the elastic potential is also an important pedagogical tool for the teaching of mechanics, acoustics and quantum mechanics. It has one fundamental property that no other interaction exhibits: It leads to periodic motions whose frequency $\nu = 2\pi\sqrt{(k/m)}$ is *independent* of the initial conditions (therefore independent of the amplitude).

All examples so far given represent central force fields (V function of r only); the forces on each material point of an interacting pair are not only parallel but lie on the same line (in Fig. 3.2a, the line joining bodies 1 and 2). Central forces are essential for systems exhibiting periodic stability. There are interactions with noncentral forces whose fields derive from vector potentials, as happens with a pair of electric point charges in uniform motion (see below). They still satisfy (3.1) but the forces are not collinear. Table 3.2 summarizes the expressions of the interaction force for the three examples mentioned thus far, including the magnetic interaction to be discussed below.

Table 3.2. Mathematical expressions for the gravitational, electrostatic, magnetic and elastic interactions between two isolated material points, discussed in the text. Note how we have grouped the factors in the third equality.

Classical interactions	Potential expressions for point particles	Representative variables/constants		
		probe	source	space factor
(a) Gravitational	$\boldsymbol{f}_g = -\mu_p \boldsymbol{\nabla}\left(\dfrac{-G\mu_s}{r} + C\right) =$	μ_p	$G\mu_s$	$-\dfrac{\boldsymbol{r}}{r^3}$
(b) Electrostatic	$\boldsymbol{f}_E = -q_p \boldsymbol{\nabla}\left(\dfrac{K_e q_s}{r} + C\right) =$	q_p	$K_e q_s$	$\dfrac{\boldsymbol{r}}{r^3}$ $K_e = (4\pi\varepsilon_0)^{-1}$
(c) Magnetic	$\boldsymbol{f}_B = -q_p \boldsymbol{v}_p \times \boldsymbol{\nabla}\times\left(\dfrac{K_m q_s \boldsymbol{v}_s}{r} + \boldsymbol{\nabla}C\right) = q_p\boldsymbol{v}_p$	$\times K_m q_s \boldsymbol{v}_s$	$\times \dfrac{\boldsymbol{r}}{r^3}$	$K_m = \left(\dfrac{\mu_0}{4\pi}\right)$
(d) Elastic	$\boldsymbol{f}_{ela} = -\boldsymbol{\nabla}\left(\frac{1}{2}kr^2 + C\right) =$	k	$-\boldsymbol{r}$	

[†] The physical origin is, however, electromagnetic – in the neighborhood of any relative minimum of the function $V(r)$, we can expand it in the form $V(r_m + \rho) \approx V(r_m) + \frac{1}{2}k\rho^2$, which is a "local" elastic potential (in one dimension) with $k = d^2V/dr^2|_m$.

3.3 Classical Force Fields

There are a few points that need to be emphasized, crucial to the understanding of the concept of interaction in the classical domain and to the concept of information as well. Let us consider only the gravitational and electrostatic interactions which work in vacuum, leaving aside all those complex ones requiring the action of artifacts or a material medium between the interacting bodies.

1. Such interactions are bidirectional in the sense that neither of the two interacting bodies has a hierarchical ranking over the other; there is no "cause-and-effect" relationship as long as there is no external interference: no irreversibility or asymmetry, and no privileged direction of time.
2. We can view (i.e., mentally model, Sect. 5.1) the force field as the *interaction mechanism* proper capable of delivering kinetic energy to the interacting bodies or receiving energy from them (potential energy loss or gain, respectively, relation (3.3)).
3. In the case of two interacting bodies the force field is in reality a superposition of two fields of independent sources: the field of the body we have placed at the origin, and the field of the probe (we already stated that fields are additive). This does not invalidate the pertinent relations a and b in Table 3.2, however, because we have tacitly assumed that the self-field of a probe has no effect on the probe itself – which is true, but only as long as the probe is not accelerated (see below).
4. Because of 3, when we view a field as anchored at its source we must assume that this self-field was mapped with an infinitesimally small mass or charge, e.g., $g = \lim_{\mu \to 0} f/\mu$, $E = \lim_{q \to 0} f/q$,[†] in which case it is guaranteed that $f \to 0$, that the contribution of the probe to the total field at any distant point tends to zero, and that, hence, the acceleration of the source $a_\mathrm{s} \to 0$ in the interaction (regardless of its inertial mass).

As a result of all this, in the case of an actual source-probe interaction, we are tempted to consider the system as consisting of three components, the source (or "cause"), its self-field (the interaction mechanism) and a small probe (a "recipient"), and view in our mind (as Maxwell did) the following "mechanistic" chain of physical relationships:

1. a local action of the source on the immediately surrounding field;
2. a propagation of this effect through the field to distant points, determining the field configuration everywhere; and
3. a local action of the distant field on the probe.

[†] Remember that we are in the classical domain. In the quantum domain, the smallest possible electric charge is that of the so-called down-quark (in absolute value, 1/3 that of the electron or proton).

The field at a distant point thus plays the role of "local ambassador" of the source. This picture becomes more than just a metaphor when we consider time variations and realize that a field can exist and act on a probe long after its own source is gone (see below). Wow! Does all this not smell of "information"? We shake the source, the information about the shaking is broadcast throughout the field, and if a probe is "tuned in," it will react! If we want to get really subjective we can state: The *purpose* of a field is to *communicate information* about happenings at its source to any distant recipient; if this recipient understands the *meaning* of the message, it will *react* accordingly (e.g., wood shavings will not "understand" the meaning of a magnetic field message, but iron filings will and dutifully line up as mandated by the field!). Unfortunately, this is at best a pedagogically useful metaphor and at worst teleological parlance ... See Chap. 5!

To introduce time variations, we consider a probe with an electric charge q_p that is moving with constant *non*relativistic velocity v_p with respect to other moving charges. One finds that in addition to any electric field force (Table 3.2b), there will be an additional force $f_B = q_p v_p \times B(r)$ (the Lorentz force). $B(r)$ is the magnetic field vector. The magnetic field of one moving point charge q_s passing through the origin with velocity v_s derives from a vector potential in the form $B = \nabla \times A$, in which $A = K_m q_s v_s/r + \nabla C$ (law of Biot and Savart for a single charge in motion). $K_m = \mu_0/4\pi = 1.000 \times 10^{-7}\,\mathrm{kg \cdot m \cdot C^{-2}}$ is the universal constant of magnetic interaction (μ_0 is commonly called magnetic permeability of vacuum; the Coulomb, unit of electric charge in the SI system, is chosen so as to make K_m *exactly* $= 10^{-7}$); ∇C is the gradient of an arbitrary scalar field. The resulting field is, of course, not static, because its source is moving. Two interacting charges in slow uniform motion exhibit an instantaneous (nonstatic) magnetic interaction force shown in Table 3.2c (in addition to the electric forces). The fact that the magnetic force (Lorentz force) is always perpendicular to the velocity of the particle on which it acts (no work done!) has far-reaching consequences for the role of the magnetic field at the cosmic scale (plasma dynamics in stars, planetary and galactic space, black holes, etc.). *Static* magnetic fields in absence of any electric field can be obtained with stationary closed electric current distributions, with B (now truly independent of time) determined as the sum of infinitesimal contributions from all linear elements dl of current filaments δi (replace qv_s in the table by $\delta i dl$ and integrate over the current loops). For a stationary distribution of charges and currents of volume densities ρ and j, respectively,[†] the electric and magnetic fields obey the static *local* differential equations

---

[†] For a point particle located at r, ρ and j are expressed as delta functions: $\rho = q\delta(r)$, $j = q\,v\,\delta(r)$. The delta function is defined as the limit of certain functions (e.g., a Gaussian) and has the fundamental property $\int_{-\infty}^{+\infty} f(x)\delta(x-a)dx = f(a)$.

shown in Table 3.3a which can be deduced from equations a and b in Table 3.2.

Table 3.3. (a) Differential equations for the static (decoupled) electric and magnetic fields. (b) Maxwell's equations for dynamic fields, plus the equation of conservation of electric charge

(a)

Electrostatic case

Sources	Field types
$\nabla \cdot \boldsymbol{E} = \dfrac{\rho}{\epsilon_0}$	$\nabla \times \boldsymbol{E} = 0$ (conservative)
$\nabla \times \boldsymbol{B} = \mu \boldsymbol{j}$	$\nabla \cdot \boldsymbol{B} = 0$ (rotational)

(b)

Electrodynamic case

Field-source links	Field types
$\nabla \cdot \boldsymbol{E} = \dfrac{\rho}{\epsilon_0}$	$\nabla \times \boldsymbol{E} - \dfrac{\partial \boldsymbol{B}}{\partial t} = 0$ (mixed)
$\nabla \times \boldsymbol{B} - \mu_0 \epsilon_0 \dfrac{\partial \boldsymbol{E}}{\partial t} = \mu_0 \boldsymbol{j}$	$\nabla \cdot \boldsymbol{B} = 0$ (rotational)

$$\nabla \cdot \boldsymbol{j} + \frac{\partial \rho}{\partial t} = 0 \ \text{(conservation of charge)}$$

From the expression of the Lorentz force and the principle of relativity[†] one *deduces* Faraday's law ($\nabla \times \boldsymbol{E} = -\partial \boldsymbol{B}/\partial t$), which couples both fields together. If on the other hand, we have the situation of a time-dependent charge density with *open* currents (e.g., a discharging capacitor), the conservation of electric charge (one of the most "absolute" conservation principles!) requires that $\nabla \boldsymbol{j} = -\partial \rho/\partial t \neq 0$. But according to the fourth equation in Table 3.3a, $\nabla \boldsymbol{j} = \nabla(\nabla \times \boldsymbol{B}/\mu_0) \equiv 0$, so the source equation for $\boldsymbol{B}$ needs a correction for the case of a time-dependent charge density (and, therefore, a time-dependent electric field). Maxwell accomplished this by making the assumption that at any point (in vacuum) where the electric field is changing (as happens in the gap between two discharging capacitor plates) an *equivalent* current density $\boldsymbol{j}_\mathrm{D}$ called "displacement current" (with all

---

[†] This principle states that it is impossible to determine if an inertial frame of reference is in uniform motion with respect to another inertial frame, by conducting experiments exclusively *inside* the first system (i.e., without interacting with things in the second system). For instance, if I move a closed circuit through a static magnetic field, I get an induction e.m.f. in the metal due to the action of the Lorentz force on the free electrons in the metal. But to comply with the principle of relativity, I *must* get that same e.m.f. if my circuit is at rest and what is moving is the magnet (in which case no Lorentz force is acting on the electrons) – therefore an *induced* electric field (free of sources) must exist (Faraday's "law"), whose action replaces that of the Lorentz force in the other situation.

the effects of a current but not representing any moving charges) appears as a new source of magnetic field. The source equation for $\boldsymbol{B}$ now reads $\boldsymbol{\nabla} \times \boldsymbol{B} = \mu_0(\boldsymbol{j} + \boldsymbol{j}_\mathrm{D})$. With this assumption and the equation of conservation of charge, $\boldsymbol{\nabla}\boldsymbol{j}_\mathrm{D} = -\mu_0\boldsymbol{\nabla}\boldsymbol{j} = \partial(\varepsilon_0\boldsymbol{\nabla}\boldsymbol{E})/\partial t = \boldsymbol{\nabla}(\varepsilon_0\partial\boldsymbol{E}/\partial t)$ at all times and points in space, so that, necessarily, $\boldsymbol{j}_\mathrm{D} = \varepsilon_0\partial\boldsymbol{E}/\partial t$. The resulting set are *Maxwell's equations*, which together with that of the conservation of charge are shown in Table 3.3b. The fascinating thing is that these equations have survived intact all great revolutions in physics: the theory of special relativity (they are intrinsically Lorentz-invariant), general relativity (in a gravitational field, a light beam is bent if $\boldsymbol{g} \perp \boldsymbol{c}$ and its frequency, i.e., the energy $h\nu$ of the photons (2.3b) changes if $\boldsymbol{g} \parallel \boldsymbol{c}$) and quantum mechanics (electromagnetic wave packets satisfy a Heisenberg-type uncertainty relation in *both* domains, classical and quantum, because $\hbar$ drops out from relations 2.1 when applied to photons – see also Sect. 3.4). No wonder that Boltzmann, in what must have been some mysterious presage of future developments, once borrowed from Goethe's *Faust* to portray Maxwell's equations: *"War es ein Gott der diese Zeichen schrieb?"*[†]

Let us look at Maxwell's equations from the point of view of algorithmic information (Sect. 1.4). They tell us what happens at any given time anywhere in space (in the entire electromagnetic field – a huge amount of information!) as a function of the spatial configuration of the sources, normally requiring much less algorithmic information for their description. But careful: The connection between field and its sources is not instantaneous. Integration of the equations tells us that the electromagnetic field at a given point and time t depends on the *history* of the sources ρ and $\boldsymbol{j}$ in a very specific way: Each element of volume of source material located at $\boldsymbol{r}_\mathrm{s}$ contributes to the field at a distant point $\boldsymbol{r}_\mathrm{p}$ with source values ρ and $\boldsymbol{j}$ that happened there at a *retarded* time $t_\mathrm{R} = t - \Delta t$, where $\Delta t = (\varepsilon_0\mu_0)^{1/2}|\boldsymbol{r}_\mathrm{p} - \boldsymbol{r}_\mathrm{s}|$. The factor $(\varepsilon_0\mu_0)^{1/2}$ is equal to $1/c$, where $c = 3\times10^8$ m/s is the speed of electromagnetic waves in vacuum.[‡] We are tempted to picture all this in our mind as a spherical wave imploding from infinity, collecting information on all the sources it encounters on its way toward the point in question, and delivering it there to configure the field at the appointed time t. For instance, when we look at a Hubble telescope image, we are looking at a deposit of information collected during the travel of a contracting wave that originated during the

---

[†] *"Was it a god who wrote these signs?"*, in Ludwig Boltzmann, *Vorlesungen über Maxwells Theorie* (available on the Internet).

[‡] In reality the electric and magnetic fields are six independent components of an antisymmetric tensor in the 4-dimensional space-time. In the relativistic formulation of electromagnetism (for which no fundamental changes are necessary except for the mathematical representation in 4D space-time) only one universal constant, the speed c of electromagnetic waves in vacuum, is used. In that case the units of $\boldsymbol{E}$ and $\boldsymbol{B}$ are the same, but the electric charge has a derived unit, dependent on the units chosen for space, time and mass.

earliest times of the Universe and collapsed into the telescope at the time of exposure. A more familiar situation arises with the sound field (an elastic force field that requires a material medium to propagate): What we hear at any given instant of time t is not what is being emitted from all acoustic sources at that very moment, but the sum of acoustical signals collected by an imploding wave at distant points at the *retarded* times $t - |\boldsymbol{r}_p - \boldsymbol{r}_s|/v_a$, where v_a is the speed of sound ("only" $\sim 330\,\mathrm{m/s}$ in air under normal conditions).

As a matter of fact, Maxwell's equations show how electromagnetic energy can be carried from one place to another in vacuum, for instance, from the source to the probe; the electromagnetic energy flux through a closed surface is given by the surface integral of the Poynting vector $\boldsymbol{S} = 1/\mu_0 \boldsymbol{E} \times \boldsymbol{B}$. Earlier we said that electromagnetic interactions between two charged bodies are bidirectional, that there is no "cause-and-effect" relationship involved – but now both bodies are disconnected time-wise: Shaking one will have an effect on the other one *later*. However, notice that when we said earlier that there was no cause-and effect relationship, we added the statement: "As long as there is *no external interference*," which meant that as long as both interacting bodies remain isolated from the rest. This is still true: Two charged bodies *left to themselves* do interact in a totally bidirectional way; only when an external *non*electromagnetic action intervenes, such as an external force "shaking" one of the charges, does this symmetry of interaction break down, and a distant cause (the *external* action) and subsequent effect (disturbance propagating throughout the field) will be present. Another important point: Whereas it is possible to have an electromagnetic radiation field at time t without any of its sources (charges and/or currents) being present, they *must* have existed at some prior time.[†]

A more detailed, albeit qualitative, discussion of an example is in order. Fig. 3.3 shows two small spheres, electrically charged with opposite polarity. Initially, everything is stationary, there is no magnetic field, and the

---

[†] Maxwell's equations (Table 3.3b) do admit a nonzero solution for $\rho \equiv 0$ and $\boldsymbol{j} \equiv 0$ at all times. They are plane waves running from $+\infty$ to $-\infty$. Strictly speaking, however, such a solution is unphysical in the macroscopic world, just as are the so-called *advanced potentials* (taken at times $t + \Delta t$ – see previous paragraph), which represent a genuine solution of the field equations, but are summarily dismissed as nonphysical (no information "back from the future"!). There is often a "chicken-and-egg-type" discussion about which is the primary physical entity, the electromagnetic field or its material sources ρ and $\boldsymbol{j}$. This issue is particularly contentious in collisionless plasma physics, where the movement of the ions and electrons is governed collectively by their own macroscopic (i.e., average) field; only when there is a contribution from an externally applied field that exerts an *external* force on each source, can one identify clearly "which comes first." An argument in favor of the field sources is that in the elementary particle realm, they can exhibit interactions of more than one kind (i.e., be controlled by more than one field).

electric field far away from the spheres is dipolar. At $t = 0$ a spark flies between both spheres and discharges them. Let us assume that the discharge process takes a very short time Δt and consider the field configuration at some time t (Fig. 3.3a). We are tempted to say that points outside a sphere of radius $r = c(t + \Delta t)$ have not yet received any information about the discharge (their corresponding retarded times at the source precede the discharge); thus, there is no difference in field configuration from that of a static dipolar field. Points within the sphere $r = ct$ have all received the message that the spheres are discharged, and the field will be zero everywhere inside. In the region between those spherical wave surfaces a complete rearrangement of fields occurs, dictated by Maxwell's equations. It is not difficult to qualitatively guess the configuration. Since there are no local electric charges present at time t, $\nabla E = 0$ everywhere and the electric field lines must all be closed: The ones from "outside" must close near the outer spherical shell, as schematically shown in the figure. Since those closing lines are moving out, in the fixed frame of reference they represent a local electric field pulse, i.e., an upward displacement current $\varepsilon_0 \partial E/\partial t$ that will generate an outer toroidal magnetic field as sketched in the figure (outer ring). But since there are no real electric current sources (moving charges) in that transition region either, this magnetic field loop must be compensated by a reverse loop as shown (inner ring) in order to leave no trace of a magnetic field on either side of the propagating transition region. The only way such a double toroidal magnetic loop can arise is to have an additional rotational electric field configuration as sketched, with the closed loops fully inside the transition region. The physical origin in the so-called near-field region of the two outward moving and oppositely directed magnetic loops is, during the brief interval of discharge, the downward (real) discharge current i between the spheres and the upward (equivalent) displacement currents j_D of the collapsing electric field (see Fig. 3.3b). The outward propagating transition region represents an electromagnetic shock wave with transport of energy; the corresponding Poynting vectors are sketched in the figure (note that this whole system works like a dipole antenna!). This shock wave literally collects the original local electric field energy density and sweeps it away to infinity, leaving behind a field-free space. We are tempted to say, quite generally, that whenever two opposite electric charges annihilate each other, electromagnetic energy is emitted as a messenger informing all points of space that "the game is over." This latter statement, however, is *not* physics: It is a description of *our subjective view,* a mental image, of the process, which stems from the fact that a field is a model closely tied to our perception of an elastic material medium (Sect. 5.1)!

Another example that can be analyzed in a similar qualitative way is that of an electrically charged body in uniform rectilinear motion that comes to a sudden stop. A spherical wave (in this case called Bremsstrahlung) will propagate from the point of impact, rearranging the field of the charge. Outside the instantaneous radius $r = ct$ lie points that have not yet "learned" of the

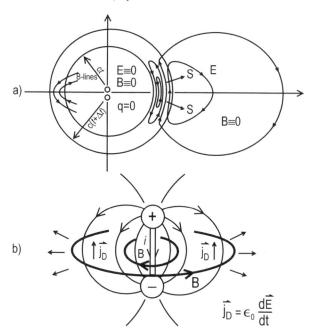

Fig. 3.3. Sketch of the electromagnetic field of two oppositely charged spheres (not in scale). (**a**) Electric and magnetic fields at time t after a spark of short duration Δt has discharged both spheres. The "field rearrangement" region is an expanding spherical electromagnetic wave represented by a source-free interplay between time-varying $\boldsymbol{E}$ and $\boldsymbol{B}$ fields. The Poynting vector $\boldsymbol{S}$ represents the energy being swept away by the wave – this is a statement describing the *physics* of the process. In common parlance, "the wave carries information about the cancellation of both charges" – this, however, is a statement describing our own *intuitive view* of the process. (**b**) Generation of the magnetic field in the "near-field" region, close to the discharge

impact (i.e., for which everything happens as if the charge had continued with its uniform motion *beyond* the point of impact); inside there will be the electrostatic field of a charge at rest. The transition region will contain coupled $\boldsymbol{E}$ and $\boldsymbol{B}$ fields that represent an outflow of electromagnetic energy. Quite generally, Maxwell's equations show that any *accelerated* electric charge will emit an electromagnetic wave, and that, conversely, a pure electromagnetic wave, under appropriate field-geometric conditions, can impart a mechanical impulse to a charged particle.

Let us return to the ideal case of two electric point charges interacting with each other in absence of any external influence (a closed system). The interaction forces are mutual (no hierarchical order, no real cause-and-effect direction), and the bodies will interchange energy with the field and, through it, with each other. But in view of the examples discussed above, there will another participant here: the electromagnetic *energy* that may be carried away by the field, because both bodies will in general be accelerated (relation (3.1)). This is why during the second half of the 19th century it was difficult to understand how in the first atomic models the electrons could stay in orbit without slowing down and falling into the nucleus. Of course, now we have reached the quantum domain and the two interacting charged particles, say an electron and an ion, must be treated as *one* system (see Sect. 2.8). That system has discrete energy eigenstates or energy levels; when it changes from one state to another it will indeed emit (or absorb) electromagnetic energy – and this electromagnetic energy is, of course, quantized in the form of a photon (relation (2.11)), the quantum equivalent of an "electromagnetic field rearrangement wave." But as long as the system remains in an eigenstate (or superposition thereof) free of external perturbations, it will remain stable (an "attractor" structure, as we called it in Table 3.1). Another example is the bound system of an electron–positron pair (positronium). It, too has discrete stable energy levels, but it comes with a built-in instability because the electron and the positron will eventually annihilate each other (overlap of their wave functions), creating a pair of energetic gamma rays (two photons of 511 keV each). This process is the quantum equivalent of Fig. 3.3, with one addendum: The entire rest energy $2m_e c^2$ of the positron–electron pair is available to the photons (and there also would be a tiny little bitty of gravitational radiation signaling the disappearance of mass!).

3.4 Quantum Interactions and Fields

To discuss qualitatively the concept of interactions and force fields in the quantum domain, we turn to our workhorse example of Fig. 3.3 and, as we did in Sect. 2.2, decrease gradually the size, mutual distance and total initial source charges until we reach down into the quantum domain. The "field rearrangement or energy-sweeping wave" will be quantized, represented by individual photons (at minimum, pairs of oppositely directed photons, to uphold conservation of momentum). Their angular distribution (probability of emission as a function of direction with respect to the axis of the initial dipole) will be in total conformity with the classical distribution of radiated electromagnetic energy (correspondence principle, Sect. 2.1). The frequency (energy) of the photons is determined by the available energy (in the case of an electron–positron annihilation this would include their rest masses).

A similar process occurs with the Bremsstrahlung of a decelerated charge: The emitted radiation is quantized in a photon, whose angular distribution is in correspondence with that of the classical radiation wave and whose energy is given by the energy and momentum conservation laws. On the other hand, the photoelectric effect in which an electron acquires momentum delivered by an incoming photon and the absorption of a photon by a bound

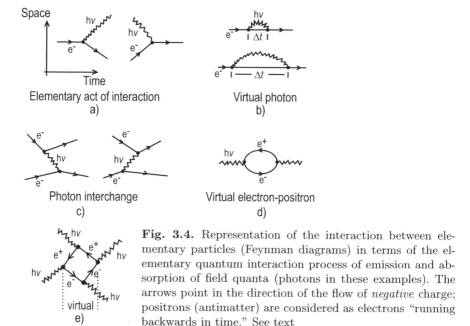

Fig. 3.4. Representation of the interaction between elementary particles (Feynman diagrams) in terms of the elementary quantum interaction process of emission and absorption of field quanta (photons in these examples). The arrows point in the direction of the flow of *negative* charge; positrons (antimatter) are considered as electrons "running backwards in time." See text

electron in an atom are examples of radiation absorption. In *quantum electrodynamics* (QED) the fundamental process is one of emission or absorption of a photon by a charged particle; this is what we could call "the elementary act of quantum interaction." It is represented with a so-called Feynman diagram [39], as shown in Fig. 3.4a, characterized by a vertex of concurring world lines governed by a coupling constant, in this case, the electrostatic interaction constant times the electric charge of the particle in question (the product $K_e q_s$ in equation b of Table 3.2). All electromagnetic interactions are described as a sum of interlinked elementary acts of interaction.

The examples I have given above involve charged particles which are interacting with some other particle or particles (the two charge components of a dipole; the electron being suddenly slowed down when it hits the atoms of a target; a bound electron in an atom). Elementary interactions between *free* charged particles and photon(s) (represented by a *single-node* Feynman diagram) are strictly prohibited; this includes electron–positron creation by an energetic photon in vacuum. Look at the left graph in Fig. 3.4a: We could describe this process acting in isolation by moving to another inertial frame which is attached to the incoming electron; in that frame of reference, the electron is at rest. After the emission of a photon, in that system of reference we would end up with a moving electron *plus* a photon – this would be a blatant violation of the conservation of energy! Likewise, for the right-side graph in Fig. 3.4a, in a frame now moving with the outgoing electron, that particle would be at rest; in that system, the initial state was an electron moving to the left plus a photon incident from the left – where did all this energy go after the photon was absorbed and the electron stopped? Again, it would be a blatant violation of conservation of energy. What is not prohibited, however, is the elementary interaction of a free charge with a *virtual photon,* one that lasts such a short time that, to reveal it experimentally, a time interval longer than the duration of its actual existence would be required.

This is indeed how QED describes a *static* electric or magnetic field: a site populated by *virtual photons* through processes of emission and absorption by the source charge (or magnetic dipole). According to Heisenberg's uncertainty principle (see relations (2.1b) and (2.3a)) the emission and reabsorption of a photon of energy $h\nu$ by a charged particle (Fig. 3.4b) is allowed as long as the time Δt the photon is "roaming around" is shorter than $\approx 1/\nu$ (notice that Planck's constant h is gone!). Since a photon is massless, the energy of a virtual photon can be as low as one wished, which means that the range $c\Delta t$ a virtual photon would be allowed to travel can be arbitrarily large, as long as it is of sufficiently low frequency (long wavelength). The electrostatic field of a charged particle at rest is thus a region of space surrounding the charge "populated" by virtual photons – the farther away, the lower their energy; high energy photons are only allowed to stay very close to the source charge. None of them is allowed to appear directly as a "real" photon in a measurement unless the missing energy is provided by an additional interaction. Yet these virtual photons do represent energy density in space: the energy density of the electrostatic field! In other words, in classical electrodynamics, Maxwell created a "mechanistic picture" in which the electromagnetic field in empty space bore properties similar to those of an elastic medium, with energy storage and momentum transport capability. In QED the picture is no different, except that field energy and momentum are embodied in a "cloud" of virtual photons surrounding the source charge – on occasion, when sufficient energy is provided from outside, they become real and take off as electromagnetic radiation.

Virtual photons and their dynamic contribution can be revealed indirectly in the case of an interaction between two charged quantum particles. Figure 3.4c shows two cases of the simplest QED description of an elementary electromagnetic interaction process with two vertices of elementary quantum interaction, in which one virtual photon is *exchanged* between the two particles. The interaction force (which is not virtual but "real," i.e., measurable) can be thought of as a "recoil" (in the case of repelling charges), or a "lasso" (attracting charges), mediated by the interchanged virtual photon. The farther away both charges are, the softer will be the exchanged virtual photons, and the weaker the force.[†] Note that if the photon had a finite rest

---

[†] This allows the introduction of a dimensionless quantity (well known from optical spectroscopy), the *fine structure constant,* as an expression of the "intensity" of the elementary electron–photon interaction (the nodes in Fig. 3.4). The potential energy of the two-electron system for that configuration is $\Delta E_e = e^2/(4\pi\varepsilon_0 r)$.

mass (which fortunately it does not – we would not be around!), the range of electromagnetic forces would be greatly limited – indeed, the slow $1/r^2$ decrease of the electrostatic field is a confirmation of the null rest mass of the photon! The fact that electromagnetic interaction forces are amenable to measurement does not invalidate Heisenberg's principle: The field of a quantum charge reveals the presence of a distribution of virtual photons, but not that of *one particular* photon. As a matter of fact, to calculate electromagnetic interaction forces in QED, it is necessary to sum over all (infinite) possibilities of a single or multiple photon exchange between two charged particles (sum over all topologically possible Feynman diagrams).

The scattering of an electromagnetic wave by another wave in vacuum is not possible in classical electrodynamics: The fields of different sources superpose linearly and propagate independently without bothering each other (this is *not* true inside a material medium). But QED allows photon–photon scattering as a higher order process. Just as an electron can emit and reabsorb virtual photons, a photon can create a virtual electron–positron pair (particle–antiparticle) during such a short time that the presence of the pair could not be revealed directly by a measurement (Fig. 3.4d). Now we can envision a process like the one shown in Fig. 3.4e, in which a second incoming photon interacts, for instance, with the virtual positron, and the virtual electron emits a second outgoing photon. This represents a scattering (an interaction) between two photons – it is a higher order process involving four vertices (its probability or cross section is proportional to $K_e^2 q_s^4$ and therefore very rare). The parametric down conversion of a photon, mentioned in Sect. 2.8, in which an incoming photon splits into two, is another example for photon–photon interaction, although it does require a special medium to occur.

Similar considerations of field quantization can be applied to the gravitational interaction, with a key quantitative difference: The small value of the interaction constant vis-à-vis that of electrostatic interactions leads to a great difficulty in observing gravitational radiation fields and their quanta, the massless spin-2 *gravitons* (massless, because the field decreases like r^{-2}, as in the electrostatic case). However, the gravitational field does play a special role. The principle of equivalence (see footnote † on page 96) demands that photons be subjected to gravitational acceleration despite being strictly massless. This means that light is bent in a gravitational field. For a similar reason, when an electrically charged particle is accelerated by a gravitational field, no electromagnetic radiation should be involved (otherwise we could

---

On the other hand, if r is the distance between two interacting electrons (see diagram c of the figure), the virtual photon must have an energy less than $\Delta E\gamma = \hbar c/r$, otherwise it would not be virtual (relation (2.1b)); r/c is the time it takes the virtual photon to go from one particle to the other. The ratio of the two, $\Delta E_e/\Delta E\gamma = e^2/(4\pi\varepsilon_0\hbar c) = 1/137$ is the dimensionless fundamental constant of electromagnetic interactions.

tell the difference between an accelerated frame of reference and an inertial one subjected to a gravitational field – yet this matter is still under active investigation). But this has nothing to do with quantum interactions – photons are not coupled to gravitons; rather, it is related to the properties of spacetime itself, whose geometry (curvature), according to the theory of general relativity, is linked to the gravitational field itself.

A little exercise is in order. Suppose we have a supermassive spherical celestial body of mass M with a very small radius (e.g., a collapsed supernova or neutron star, but not rotating in our hypothetical example). If we inject a photon perpendicular to the radius vector r, a purely classical calculation tells us that the radius of curvature of its trajectory would be $\rho = c^2/g(r)$, where $g(r) = GM/r^2$ is the body's gravitational field at the injection point. For $\rho = r$, the "orbit" of the photon would be circular. General relativity introduces corrections to these relations and defines the Schwarzschild radius $\rho_s = 2GM/c^2$ (which plays a fundamental role in black hole physics) for which, indeed, a photon would be in circular orbit around a massive central object. Since $G/c^2 \cong 10^{-27}$ m/kg, it will have to be an extremely massive and at the same time very small object to obtain an orbital radius reaching beyond the body's surface. Suppose now that you have a *point* particle of inertial/gravitational mass M (assume zero charge and spin). How close to this point-like particle will a photon have to be to be trapped around it? Given the extreme smallness of the factor G/c^2, any "normal" particle mass will give a radius that is well below any imaginable limit – certainly below any limit set by the Heisenberg relation for measurability. If $\Delta t \approx \rho_s/c$ is the time scale of the "photon in orbit," according to relation (2.1b) we must have $\Delta t \Delta E \geq \hbar/2$, in which ΔE should be energy of the system, roughly $\approx Mc^2$, the rest mass of the particle. This gives a lower limit of the order of $M_P = (\hbar c/G)^{1/2} = 2.177 \times 10^{-8}$ kg, called the *Planck mass*. The value $M_P c^2 \approx 1.2 \times 10^{19}$ GeV would be an enormous mass for a single elementary particle! From this mass one derives the *Planck time* $t_P = \Delta t = \rho_s/c = (\hbar/Gc^5)^{1/2} \approx 5.391 \times 10^{-44}$ s, and the *Planck length* $l_P = \rho_S = (\hbar/Gc^3)^{1/2} \approx 1.616 \times 10^{-35}$ m, mentioned in Sect. 3.1. The more rigorous definition (yet leading to the same expression) of the Planck length is given in footnote † on page 82. For a general discussion of these domains, see [4, 50].

We already mentioned the existence of other types of interactions (see Table 3.1). We have strong interactions between quarks (and through them, between nucleons inside a nucleus), whose field quanta are the *gluons,* and electroweak interactions between quarks, mediated by three *gauge bosons,* the massive $W^\pm$ and Z_0 particles – so massive that the electroweak force is very short-ranged. The strong interaction, also short-range, is reversible and there is no privileged direction of cause-and-effect or time. As mentioned before, for each fundamental interaction mechanism there is a universal coupling constant; their actual values (and the relation to the universal constants of gravitational and electromagnetic interactions) determine in a remarkably sensitive way the formation and stable existence of the entities shown in the third column of Table 3.1. Recent high-energy experiments point to the fact that the coupling constants are energy-dependent, and that somewhere around 10^{16} GeV the values of the electromagnetic, weak and strong interaction constants should coincide (e.g., see [31]). It is interesting to note that small changes, particularly in the strong interaction force, could still lead to an evolving Universe, but the formation of stable stars and the nucleosynthesis of carbon – hence the existence of planets and life as we know it – would be impossible!

Because of their role as interchange particles, virtual field quanta (photons, gluons and other gauge bosons) are sometimes called *the messengers* in elementary particle interaction processes. The field quantum is "sent" by the source particle to the probe, "causing" it to experience a force. This process could be described in even more anthropomorphic terms: "Hey, here is an intruder (the probe particle) – let's kick him out or bring him in" (depending on whether the interaction is repulsive or attractive). The classical field "ambassador" mentioned earlier in connection with the electromagnetic field has been replaced by a "messenger." But careful: An electron does not "send out" one *particular* virtual photon exactly toward the probe with the *purpose* to interact with it! Indeed, there is not just one such virtual particle – it is a stochastic process with a whole "cauldron" of unobservable entities around the source!

We have been throwing around several times the word "information." We made statements like: "The field at a point acts as the 'ambassador' of the source"; "a wave collects information on all sources that it finds on its way"; "points outside a certain sphere have not learned of the impact"; "the field quanta are messengers"; etc. What kind of information is this? It certainly is *not* something that makes the Universe tick – forces, fields and energy do that. Every time we have alluded to information thus far, we were dealing with information *for us*, for our models, for our description, prediction and retrodiction of a physical system. In quantitative descriptions it is mainly algorithmic information on the features of a physical process such as the field-reorganization wave of Fig. 3.3, the Bremsstrahlung wave emitted by electrons hitting the target in an X-ray apparatus, or the photons in the Feynman diagrams of Fig. 3.4. Such information, indeed, requires less bits than a full point-by-point and instant-by-instant description of the entire system, and thus conforms to the definition of algorithmic information given in Sect. 1.4. Overall, the use of the word information in the preceding paragraphs was purely metaphorical: It was used as an instrument in our attempts to comprehend the physical universe in terms familiar to our biological brain. We shall return to this discussion in Chap. 5.

3.5 Information-Driven Interactions and Pragmatic Information

When we *use* an electric discharge like that of Fig. 3.3 as a *sender* of information (a transmission antenna!) with the *purpose* to elicit some specific *change* (response) in a distant *recipient,* the entire electromagnetic system becomes part of a genuine information system, representing what we shall define below as an *information-driven interaction* between a sender and a recipient. But let me emphasize: When the electromagnetic wave of Fig. 3.3 is the result of a natural discharge like an atmospheric lightning strike, no information whatsoever is involved – the wave represents field reorganization and energy propagation – nothing more, nothing less. In this section we shall examine in depth, and define, the concept of information-driven interactions.

Let us consider a complex system, consisting of many interacting parts – say, billiard balls on a frictionless table. Each interaction between two balls is perfectly elastic, i.e., reversible. Consider the situation shown in Fig. 3.5. We have an "input layer" or "detector" I of balls and an "output layer" or "effecter" F. At an initial time the balls {A, B, C, ... } of the input layer have a given distribution of velocities V_A, V_B, etc., triggering a chain of elastic collisions that finally impart velocities V_M, V_N, etc., to balls {M, N, R, ... } of the output layer F. Hamiltonian mechanics allows us to calculate what the final state of the output layer balls will be (and of course that of all other balls, too – see Sect. 5.2). Now assume that all intermediate balls are in a "black box" to which we have no access. In that case we can consider the entire process as an input system I coupled to an output system F by the action of *one* complex interaction mechanism, the balls in the black box. The interaction is not only complex but also *irreversible* because, to occur in an exactly reverse way so that each ball in I flies off with exactly the same speed but in reverse direction, it will not be enough that {M, N, R, ... } start with reversed velocities $-V_M$, $-V_N$, etc.: *Many* other conditions must be fulfilled exactly (e.g., each collision partner inside the "black box" must arrive at the right place and time with the right (reversed) velocity). There is energy coupling between the balls of the input layer and those of the output layer, but it will be weaker the more complex the interaction mechanism (the more intervening balls there are).

Suppose we want to assure a univocal relationship between the input state and the final output state so that for a given pattern of initial velocities of the balls in I the corresponding pattern of outflying balls of F is always the same. The only way to assure this is to add some extra device that *resets* all balls in the black box to exactly the same initial positions (at rest). In other words, if this interaction mechanism is given the *purpose* to always elicit the same output for a given input pattern, it must have the ability to *reset itself* to the same internal state every time – otherwise the desired correspondence would not occur, or just happen by chance. As the interaction mechanism and the input and output patterns become more complex, the

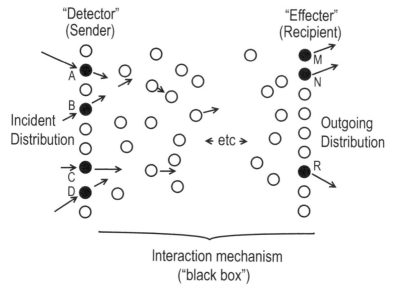

Fig. 3.5. Example of a complex physical (force-based) interaction between a "detector" of incoming impulses and an "effecter." Initially, all balls are at rest. At each step there is transfer of energy, but the amount of energy transferred between input and output could be very small if the interaction mechanism is very complex and no extra energy is fed into the system. For each *given* pattern of initial impulses, there is *one* specific outcome (pattern of outflying balls). If this interaction is given the *purpose* of consistently eliciting the same output for a given input pattern, the interaction mechanism must have the ability to *reset itself* to the same internal state every time. The energy relation between input and output is irrelevant; indeed, many input constellations may be totally equivalent energywise yet trigger very different outgoing distributions. An interaction under these particular conditions is called an *information-based* interaction; the input layer is the "sender" and the output layer a "recipient." In any natural (not artificial) system with similar input-output properties, the interaction mechanism could only emerge through biological evolution, or, in the neural system, also through a learning process (see text). From *Roederer* [96]

resetting process would require more and more extra operations. Note that since the mechanism remains the same or goes back to its initial state every time it is activated, the interaction process has a catalytic character. The purpose itself represents information and requires memory (see later, when we discuss neural networks, Chap. 4).

Notice that the actual correspondence between input and output patterns is governed by the configuration of the initial positions of the balls inside the interaction mechanism. Changing the memory in the resetting process that keeps a record of these initial positions will change the interaction between I and F, hence, the output for a given input. Suppose now that we want to *specify* the output distribution for a given input. In principle, we could calculate the initial positions of the balls in the black box (the interaction mechanism) that would ensure that new correspondence. Moreover, for a sufficiently complex system, we could specify the output distributions for a whole *set* of input constellations and calculate the configuration to which the balls in the black box would have to be reset in order to assure univocal pairs of input-output correspondences. In Sect. 4.2 we will hint how this could be done, but here we ask ourselves: How would *Nature* do it? Obviously, it would require a trial and error process. Given a large number of identical copies of the system with the same initial setup of balls in the black box and some process which introduces small random variations ("errors") in the reset mechanism, only those systems which *evolve* in the right direction, i.e., whose end states come closer and closer to the wanted values, would be able to survive while those less fit would disappear. A system like Fig. 3.5 of course does not exist in Nature, but it can serve as a metaphor for two important cases: neural networks, with the synaptic connections playing the role of the initial positions (Sect. 4.2), and the genome (Sect. 4.4), in which the positions of the bases in the DNA play the role of the initial positions of the balls (no resetting necessary here, because DNA-mediated interaction processes are catalytic in the sense that the DNA emerges reconstituted from the entire process).

Let us point out some fundamental characteristics of a system like that of Fig. 3.5 with a reset capability: 1. a complex interaction mechanism; 2. the existence of a purpose (to obtain a given result every time, not just by chance); 3. a univocal correspondence between patterns at the "detector" or sender and the changes at the "effecter" or recipient; 4. no direct energy coupling between detector and effecter (although energy is necessary for the entire operation). We call this class of interactions *information-based* or *information-driven* [95], and distinguish them from the force-driven interactions of all natural inanimate systems. Careful: We have only identified and given a *name* to a certain class of interactions – we have *not* yet defined the associated concept of "information."

It is important to discuss now some more realistic prototype examples with similar properties and compare them with homologous cases of force-driven interactions. Consider the four cases, all in the classical domain, sketched in Fig. 3.6. The first example Fig. 3.6a shows an electrically charged particle passing by a fixed (e.g., massive) charged body – a purely force-driven interaction. For each given initial condition of position and velocity, there will be a well-defined orbit. At all times, there is energy give-and-take between the particle and the field of the central body. If both bodies are, say, components of an ionized gas, or if they are electrically charged dust particles in a planetary ring, no "purpose" could be identified for such an interaction. The second example Fig. 3.6b represents the trajectory of an insect flying toward a light source. Just like the charged particle in Fig. 3.6a, an interac-

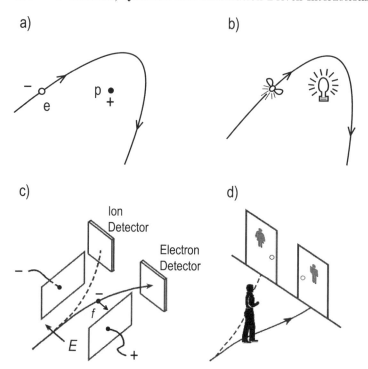

Fig. 3.6. Examples of familiar force-driven and information-driven interactions. See text for details

tion between the two bodies (insect and light source) leads to a well-defined trajectory – but in this case the interaction mechanism is complex. It does involve energy and forces but there is no direct energy balance between the two interacting bodies: The insect's wings provide lift and propulsion, i.e., the force that determines the orbit; the insect's metabolism provides the necessary energy. But it is the two-dimensional angular *pattern* of light, not the flux of electromagnetic energy from the source per se, which through the insect's sensory-motor system regulates the force. Here we have an information-driven interaction with a clear biological purpose on the recipient's side. The interaction mechanism, i.e., detection of light pattern, is unidirectional, attached to the recipient. In some species it can be rather simple [19]: It automatically causes the amplitude of the wing flap to decrease on that side of the animal which faces the light; this leads to a self-correcting turn of direction. It is obvious that such a mechanism is inherited, the result of evolution.

Example Fig. 3.6c shows an oversimplified electrostatic lens in a particle spectrometer. Again, this involves a physical interaction between the incoming charged particle and an electrostatic field, but with one big difference with case Fig. 3.6a: This is a human-built artifact whose function is to establish a correspondence between an input pattern (mass, charge or angular distribution of the incoming particles) and the angular distribution at the output (so that the particles can be separated into different categories and counted). In other words, this is a purely force-driven mechanism but with a *purpose* to achieve something that would not happen naturally, or just happen by chance. Later, in Sect. 5.2, we will classify it under information-driven interactions, because it is the result of an intentional human action (see also below). Finally, the fourth example Fig. 3.6d is rather similar to Fig. 3.6b; it has a clear purpose, but in this case, the biological interaction mechanism involves truly complex processes and decision-making in a person's brain. On the basis of inherited neural hardware and *learned* operations, it clearly reveals the existence of a "common code" between the sender and the person – a purpose on part of the sender and an understanding of the symbols on part of the recipient (in case Fig. 3.6b there is a common code, too, but it evolved genetically). This is also a good example to show how information (conveyed by the door symbols) serves to "tip the balance" into one consistent direction (predictability: always going through one door and not the other – a binary alternative with zero Shannon information!). In effect, the man's brain is initially in a quasistable symmetric state (seeing two restroom doors – a Cbit (Sect. 1.5)) until it processes the key information-bearing optical signals (see also Table 1.1).

In the Introduction we mentioned that information can be extracted from, or deposited in, the environment. Without as yet defining information, let us analyze these processes from the point of view of interactions. Take the one mentioned in the Introduction: a person walking through a forest. The person responds to the visual perception of the obstacles, a complex process that involves information-processing and decision-making at the recipient's end with a definite purpose. At the obstacle's (the sender's) side, we have scattering and/or reflection of incident light waves; no information and purpose are involved – only physical processes are at work here. There is no energy coupling between sender and recipient; again, what counts is *not* the energy of the electromagnetic waves but the *pattern* of their 2D angular distribution (determined by features of the obstacle's surface). In the recipient's side there is a purpose and meaning, but it is not given by symbols in the input (like in Fig. 3.6d), rather it is given by the physiological and neural mechanisms that are seeking out certain input patterns (like in Fig. 3.6b). This is a prototype of interactions that involve *information extraction:* It is a fundamental aspect of the interaction of any organism with the environment. An equivalent example at the cellular or molecular level would be that of a cell that has sensed the lack of a given nutrient, and is responding by starting up the synthesis of enzymes necessary to metabolize other nutrients. The signal (e.g., glucose concentration) has neither purpose nor information, but becomes information when detected by the metabolic machinery of the cell.

Finally, in physics, information extraction is the most fundamental purpose of any measurement process (Sect. 5.6).

The other example mentioned in the Introduction was that of an ant leaving a scent mark along its path. This is the reverse of the preceding example, with *information deposition* in the environment for later use. The sender has a specific purpose and information is involved. At the other end of the interaction (the path) only physical (chemical) processes are at work. This is an example of deliberate environmental modification with a purpose (another ant can extract information from the scent signals). At the cellular or molecular level we can refer to the methylation process, which changes the cell's ability to transcribe certain portions of its DNA into RNA, without changing the information content of the DNA. This alters the program of the genes that they can express, thereby rendering these cells different from the others. Methylation is heritable, and the change persists across multiple cell divisions (see also Sect. 4.4). In physics, besides the obvious example of writing a paper, a more fundamental example is that of the deliberate setting of initial or boundary conditions of a physical system in a real or a thought experiment (Sect. 5.2); case Fig. 3.6c is also an example.

To recapitulate: In information-driven interactions the responsible mechanisms always consist of physical and/or chemical processes; the key aspect, however, is the *control* by pattern-dependent operations with an ultimate purpose. There is no direct energy coupling between the interacting bodies (light bulb–insect, symbol–movement, tree–person, ant–track, physicist–initial conditions), although energy must be supplied locally for the intervening processes. Quite generally, in all natural information-based interaction mechanisms a *pattern* is the trigger of physical and chemical processes; energy or energy flows are required by the processes to unfold, but they do not govern the process. As we shall see in the next chapter, natural mechanisms responsible for this class of interactions do not arise spontaneously: They must *evolve* – in fact, Darwinian evolution itself embodies a gradual, species-specific information extraction from the environment. This is why *natural* information-driven interactions are all *biological* interactions.

Let us sharpen the concept of information-based interactions and related definitions. First of all, we note that information-based interactions occur only between bodies or, rather, between systems the complexity of which exceeds a certain, as yet undefined degree. We say that system A is in information-based interaction with system B if the configuration of A, or, more precisely, the presence of a certain spatial or temporal pattern in system A (the sender or source) causes a specific alteration in the structure or the dynamics of system B (the recipient), whose final state depends *only* on whether that particular pattern was present in A. The interaction mechanism responsible for the intervening dynamic physical processes may be an integral part of B, and/or a part of A, or separate from either. Furthermore: 1. both A and B are *decoupled energywise* (meaning that the energy needed to effect the changes in system B must come from some external sources); 2. *no lasting changes* occur as a result of this interaction in system A (which thus plays a catalytic role in the interaction process – unless the pattern is erased by a separate mechanism, or a temporal pattern is obliterated or disappears); and 3. the interaction process must be able to occur *repeatedly* in consistent manner (one-time events do not qualify) – the repetition, however, requires that the mechanism and the changes in the recipient B be *reset* to the same initial state.

We further note that primary information-based interactions are *unidirectional,* going from the source or sender to the recipient. This really means that they are "actions," despite of which I shall continue to call them *inter*actions. In part this is justified, because for the process to work, there must have been some previous relationship or "understanding" between sender and recipient. Moreover, in the case of quantum systems, information extraction involves a "feedback" perturbation of the system being measured – it is not a one-way process (e.g., Sect. 2.1). In all information-based interactions, there is an irreversible cause-and-effect relationship involved, in which the primary cause remains unchanged (the pattern in A) and the final effect is represented by a specific change elsewhere (in B). In summary, a specific one-to-one *correspondence* is established between a spatial or temporal feature or pattern in system A and a specific change triggered in system B; this correspondence depends only on the presence of the pattern in question, and will occur every time the sender and recipient are allowed to interact (see Sect. 3.6). Although the interaction mechanisms involved are usually extremely complex, the patterns we are talking about do not have to be complex entities. Indeed, a pattern could be "one-dimensional," like the intensity or magnitude of some physical variable: the concentration of a substance in the solution surrounding a bacterium, the temperature of an object in contact with the skin, the intensity of a light source, etc. A servomechanism like a thermostat is an example of a man-made device that responds to only one parameter (temperature variation).

We will now make the conscious decision to keep all man-made artifacts (e.g., Fig. 3.6c) and artificial intelligence devices (and also artificially bred organisms and clones) out of the picture. Therefore, we demand that both A and B be *natural* bodies or systems, i.e., not manufactured by an intelligent being. One further point needs clarification: We already implied in Sect. 3.2 that for a Hamiltonian system of mass points we can view the initial conditions as a given input pattern, and the positions and momenta of the mass points at a time t as an output pattern, which is in univocal correspondence with the initial pattern. In other words, we could view the process as an "interaction" between a complex system A represented by the initial state at time t_0 and a complex system B given by the state of the same system at time t. If the initial conditions are natural and not man-made (like the actual positions of the planets at time t_0), we are in presence of a purely force-driven interaction, with no set purpose, just like Fig. 3.6a (the initial conditions are initial for us, because it is us who have selected the initial time t_0 – we could have chosen any other instant and it would have made no difference for the system; we shall come back to this point in detail in Sect. 5.2). But if it is us who deliberately *set* the initial conditions of a system (even if only in a model, or in our minds), a purpose *has* been set, and we have an information-based interaction like the case of the spectrometer in Fig. 3.6c. The "mechanism" that connects input to output is purely physical; in these physical interactions information plays no role – it does so only in the initial *setup* (the artificial setting of initial conditions). In a biological system, information is at play in the interaction mechanism proper – the interaction itself *is* the purpose!

Although we have been talking about information, we have not yet provided an appropriate formal definition. We declare that *information is the agent that embodies the above-described correspondence:* It is what links the particular features of the pattern in a source system or sender A with the specific changes caused in the structure of the recipient B.[†] In other words, information represents and defines the univocal character of this correspondence; as such, it is an emergent property of the physical systems involved and thus an irreducible entity. We say that "B has received information from A" in the interaction process. It is important to realize that without a reset process "complex body B" would be left in a different state (and therefore no longer be exactly the same "complex body B"). Note that in a natural system we cannot have "information alone" detached from any interaction process past, present or future: Information is always there *for a purpose* (to evoke a specific change in the recipient that otherwise would not occur, or would happen just by chance) – if there is no purpose, it is not information. Given a complex system, structural form and order alone do not represent information – information appears only when structural form and order lead to specific change elsewhere in a consistent and reproducible manner, without involving a direct transfer or interchange of energy. Thus defined, we can speak of information only when it has both a sender and a recipient which exhibits specific changes when the information is delivered (the interaction occurs); this indeed is what is called *pragmatic information* (e.g., [64]).

Concerning the interaction mechanism per se, i.e., the physical and chemical processes that intervene between the sender's pattern and the corresponding change in the recipient, note that it does not enter explicitly in the above definition of information. However, it must assure the required uniqueness of

---

[†] This operational definition of information has been criticized as being *circular:* "*Information* is what links a pattern at the source with a change in the recipient in an *information*-driven interaction." It is not! As stated before, I have chosen the term information-driven interaction for lack of a better word – it could as well have been called "type B interaction" (with "type A interactions" being the physical ones of Sect. 3.4). This type of interactions was defined without appealing to the concept of information – the latter *emerges* from that definition.

the correspondence[†] – indeed, many different mechanisms could exist that establish the same correspondence. In the transmission from A to B information has to "ride" on something that is part of the interaction mechanism, but that something is not *the* information.

We should return to the example Fig. 3.6d. It clearly shows that "purpose" is related to the intention of the sender of eliciting a specific change in the recipient, whereas "meaning" is a concept that appears in connection with the recipient, actually eliciting the desired change. Both are needed and must somehow be matched with each other to lead to an information-based interaction. Sometimes it is stated that an "accord" or "common code" must exist between sender and recipient so that information can be conveyed and interpreted. As I shall explain later, the word "accord" used in this context is a rather anthropomorphic term designating the complex memory processes that must operate within the interaction mechanism (see Fig. 3.5 and related text) to ensure a match between purpose and meaning. Information, information-processing, purpose and meaning, structure and function, are thus tied together in one package: the process of information-based interaction.

To conclude this section, let us visit again the system of Fig. 3.3 and turn to the first paragraph of the present Sect. 3.5. We are dealing with a purely physical system that can be used as a mechanism for an information-based interaction *provided* that: 1. there is a purpose for the sender to actually trigger a discharge and initiate the propagation process; 2. there is a recipient somewhere in the system (e.g., a receiving antenna) that is adequately prepared to react when the wave passes through (to change in some specific way according to some common code – e.g., appropriate tuning to relevant electromagnetic field changes); and 3. the system can in principle be reset after use so that it can be discharged again. Entirely equivalent conditions exist for the scheme of Fig. 3.5, if it is meant to portray an information-based interaction. Now go back to quantum mechanics and Sect. 2.8, especially to Fig. 2.5 and the related discussion. We said that the collapse of the quantum wave function of an entangled system cannot be used to transmit classical information – in particular, that observer A (Alice) in Fig. 2.5 cannot transmit usable information to observer B (Bob) by just carrying out her measurement. We expressed it in quite specific words: "... the collapse cannot be used to cause an intentional macroscopic change somewhere else that otherwise would not occur or would only happen stochastically." It should be clear now why we had to add the qualifying phrase back in Sect. 2.8.

It is important to realize that different patterns in a sender may elicit identical changes in the recipient. This will depend on the particular mechanism;

[†] Biologists tend to use the term "correlation" (e.g., the "neural correlate" of a percept or of consciousness – see Chap. 6). I prefer to use the term *correspondence*, because correlation has a more restricted meaning in physics.

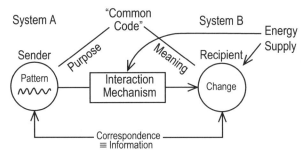

Fig. 3.7. In an information-based interaction a *correspondence* is established between a pattern in the "sender" and a specific change (structural or dynamic) in the "recipient." Information is the agent that represents this correspondence. The pattern could be a one-dimensional spatial sequence of objects (e.g., chemical radicals in a molecule), a temporal sequence (e.g., the phonemes of speech), or a spatio-temporal distribution of events (e.g., electrical impulses in a given region of the brain). The mechanism for a natural (not artificially made) information-based interaction must either emerge through evolution or develop in a learning process, because it requires a coupling between a purpose and a meaning that could not be expected to appear by chance. There is no direct energy transfer between the sender and the recipient, although energy, to be supplied externally, is involved in all intervening processes

in that case, we say that the initial patterns are equivalent and that they bear the *same* information. In other words, pragmatic information is defined by the change evoked, not the particular form of the triggering pattern. On the other hand, we will come across complex information-based interactions, in which the same sender's pattern may trigger one change from a repertoire of several possible alternatives, even if all circumstances are the same (e.g., ambiguous sensory perception, the measurement of a quantum system in a superposed state).

All the above is schematically summarized in Fig. 3.7 which points out the fundamental concepts involved. I believe that the much-sought boundary between physical and biological phenomena can be found wherever or whenever a force-driven complex interaction becomes information-driven by natural, not artificial man-made means. I would go even further and venture to define "living system" as any natural (not man-made) system capable of information-driven interactions [91] (to be capable of this type of interaction it must also be able to evolve, which requires reproduction). In the next two chapters we shall examine fundamental aspects of biology and physics in the light of the above definition of information from an objective and pragmatic point of view.

3.6 Connecting Pragmatic Information with Shannon Information

In the preceding section we identified the existence in Nature of a very special class of interactions in which the pattern in a "sender" elicits a specific, univocal change in a "recipient," and defined information as *that which represents the univocal correspondence between the pattern and the evoked change.* It is important now to establish links between the concept of pragmatic information and the Shannon information measure (Sect. 1.3). For that purpose, we turn again to our classical pinball machine, and view the output state (see Fig. 1.1) as a *pattern.* Indeed, we could wire some switches in the bins to an instrument that gives the reading on a dial with two final positions, 0 or 1, representing the evoked change after the machine has been activated (Fig. 3.8). A pointer would tell whether the machine is "ready" (when it is in a reference position, or a reset position from a previous activation), and, after the measurement, it would identify one or the other resulting state $|0\rangle$ or $|1\rangle$. The final position of the pointer (e.g., the angle it makes with respect to the reference or reset position) is the output pattern in question; obviously, there are two possible patterns in this case. In a multiple pinball machine like the one shown in Fig. 1.3a, there would be four positions of the pointer, i.e., four possible patterns, corresponding to the states 00, 01, 10, and 11. Now we establish an information-driven interaction by activating the machine and having an observer look at the result. The observer's brain is the recipient system, and the change elicited is a change in its *cognitive* state (commonly called "new knowledge," Sect. 1.1 and Table 1.1), which, as we shall discuss in Sect. 4.2 and in more detail in Chap. 6, is represented by an ultra-complex yet very specific pattern of neural activity (of hundreds of millions if not billions of neurons). And there is a one-to-one correspondence between the input pattern (position of the pointer) and the neural pattern elicited in the brain, schematically indicated in Fig. 3.8. It is important to realize that the observer (eyes and brain) can be replaced by a recording device. This does not invalidate the fact that regarding the setup shown in this figure, *a human being must have been involved* at one or another stage of planning and construction. If there is no recording device involved and no natural mechanism resets the system, no purpose could be identified and we would have a purely force-driven interaction – no information would be involved.

First, note the rather puzzling fact that, by definition, in an information-driven interaction, the expected average Shannon information content (Shannon entropy, relation (1.3)) *is always zero!* One given pattern (say the result "01") will always elicit the same pattern corresponding to the cognition of "01" (otherwise something would be wrong with the interaction mechanism – for instance, with the eyes or the brain of the observer!). We thus may venture the apparently paradoxical statement that pragmatic information always has zero Shannon information content. The reason is that pragmatic information represents, by definition, a *univocal* relationship between

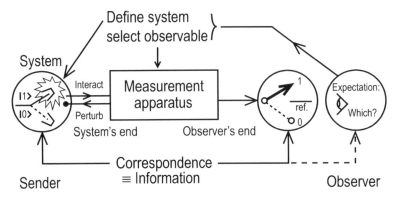

Fig. 3.8. Same as Fig. 3.7 for the case of an observer (the recipient) receiving information from a sender (the position of a pointer indicating the result of a binary pinball machine (Fig. 2.1)). The observer could be replaced by a recording device – this does not change the fact that at some stage an intelligent being must have been involved with the setup. See also Table 1.1

a source pattern and a target pattern (if the response were to be random it would mean that no "understanding" exists between sender and recipient – this is why we insist in calling this one-way cause-and-effect relationship an *inter*-action!). Either patterns could be as complex as the DNA molecular code or the face of a person, or as simple as a blip on a scope or click in a loudspeaker – let us not confuse the minimum number of bits needed to represent the *patterns per se* (this is algorithmic information, Sect. 1.4) with the existence of a causal connection between two patterns (the essence of pragmatic information). It is only when the possibility of *alternative input patterns* appears (the responses from different counters, varying intensities of a signal, etc.), that the Shannon information measure does come in!

In summary, I contend that *information-driven interactions and the related pragmatic information are the primary, fundamental concepts*. Shannon information is a derived concept, applicable to a subset of special cases – it requires previously known alternatives and knowledge of their probabilities of occurrence (Sect. 1.3). Unfortunately, we must defer to the last chapter a discussion of how the degree of uncertainty or Shannon entropy H (1.3) and the novelty value I of an actual result (1.1) can both be interpreted in terms of information-driven interactions and cognitive information processing in the brain. Let me just say at this stage that the animal brain has the capability of triggering pattern-specific neural activity by associative recall (or, in humans, imagination) in *anticipation* of an event (Sect. 4.3); when that activity is either confirmed or changed by the actual input from a measurement, the ensuing emotive reaction of the brain is linked to H and I ("what an odd thought!" some of you will say, but wait until Chap. 6!), and to the nontrivial fact that to know, you must first *want* to know (see also Table 1.1)! We are talking here about the origins of the "boring *What's new* reaction" and the "exciting *Wow* factor" mentioned in Sect. 1.2!

One clarifying note on nomenclature: In what follows, whenever I say "Shannon information" or "algorithmic information" I am talking about quantitative *measures* of information (defined in Sect. 1.3 and Sect. 1.4) regardless of what that information means. When I say "pragmatic information," or just plain "information," I am talking about the *objective concept* of information (defined in Sect. 3.5) that has meaning but which cannot be expressed with a number.

Finally, let us return to the concept of complexity (Sect. 1.4 and Sect. 3.1). The full description of a complex system (its topological and/or temporal structure) requires a certain amount of Shannon information (number of decisions, listing of pixels, etc.). If that description can be generated by an algorithm (formula, program) in fewer bits, the algorithmic information content would be less than the Shannon information content (Sect. 1.4). None of these measures, however, has anything to do with what the complex system actually portrays, its purpose, and the change, if any, that it is expected to cause in some recipient. Pragmatic information – which is what links the form and pattern of the complex system with the evoked change – is in principle unquantifiable. There is *no connection* between the degree of complexity of something and the pragmatic information it may carry. On the other hand, in a purely physical interaction of a complex system with no specific goal or purpose in a natural, nonbiological context, there is no information involved at all, as we shall discuss in Chap. 5. The only information that we humans could attach to such a complex system is Shannon or algorithmic information needed for its "geometric" or "topological" description (in space and time).

In the Introduction we asked: *Is information reducible to the laws of physics and chemistry?* Our answer now should be: *If defined in an objective way it is reducible, indeed – but only in the biological realm.*

4 The "Active" Role of Information in Biological Systems

In Sect. 3.1 we briefly described the evolution of the Universe and the formation of stable entities out of chaos by the combined effect of stochastic fluctuations and interactions governed by physical laws. Among the submicroscopic structures that emerged are the nuclei of stable isotopes of the elements and stable atoms and molecules. They assembled into "islands" of organization and order in highly localized, confined regions of space. During their formation, they remained thermodynamically open, which allowed free energy to flow in and entropy (low grade energy) to flow out. Through the effects of gravitational, electromagnetic and chemical forces (the latter also based on electromagnetic interactions), condensed matter formed in macroscopic "chunks" with clearly defined two-dimensional surface boundaries. Some of these chunks, like the crystals, exhibit high symmetry and a potential to grow and multiply under particular conditions; others, like the stars, entertain internal dynamic processes fueled by free (nuclear) energy reservoirs maintaining their structural and functional organization during a certain period of time; others yet exhibit a robust structure with high immunity to external perturbations. After their formation, none of these macroscopic structures can remain in *total* isolation because none can be shielded completely from interactions with the rest of the Universe – even if it were only for an absolutely minute gravitational field from some utterly distant celestial object [15]. And none of these inert systems, however structured and organized, can take "corrective actions" to systematically counter external effects pernicious to the integrity, stability or further organization of their structure and function. There is no purpose to do so, no mechanisms to detect fluctuations and gradual changes and discriminate among good and bad ones, and no mechanism to take any corrective action! So on a celestial scale, every macroscopic physical object will ultimately suffer a decay of its internal physical and chemical structure, and its initially low entropy density will eventually catch up with the average entropy density of the Universe. In Wagnerian German it is called "der Wärmetod aller Dinge."[†]

Yet among the organized structures in the Universe one class did emerge, based on the carbon atom chemistry, that is capable of "consistently beating the elements," i.e., maintain during a certain period of time a metastable

---

[†] The thermal death of all things.

nonequilibrium state of low-entropy and high organization in a changing and often hostile environment: the *living organisms* (so far found only on planet Earth, but suspected to develop elsewhere under similar conditions). Three main conditions have to be fulfilled concurrently to make this "consistently beating the elements" possible: 1. *encapsulation* in a protective but permeable boundary that allows for a highly selective interchange of matter and energy with the environment; 2. *self-adaptation* to environmental change; and 3. *reproduction.* The puzzling thing is that this triad, which can be summarized in reverse order as "genetics, metabolism and containment," must have developed simultaneously in tightly coordinated steps, not one after the other. Condition 2 requires the operation of mechanisms that establish correspondences between some critical spatial and temporal patterns in the environment and dynamic actions within the organism needed for metabolism, adjustment and regeneration; according to the definition given in Sect. 3.5 we are talking about the capability of entertaining *information-driven interactions* with the environment (see Fig. 3.7).

The existence of mechanisms responsible for correspondences between key environmental features and patterns of response implies that information on relevant environmental configurations (the environmental *niche*) is being acquired and stored in the living organism. While some inanimate natural bodies may be encapsulated (e.g., some geologic minerals) or may be able to multiply (e.g., prebiotic polymers), only living organisms can entertain information-driven interactions to systematically maintain a low-entropy, highly organized quasiequilibrium state despite nonperiodic, stochastic degrading influences from outside. Inert systems able to do this would necessarily be human-made, or fabricated by human-made machines. Leaving aside for a moment advanced multicellular organisms capable of acquiring information in real time through a sensory apparatus, the adaptive memory needed to sustain information-driven interactions of a living system with the environment can emerge only through a serial process in which bearers of "bad" instructions are eliminated, leaving those with "good" ones to multiply – the Darwinian selection process. To work, this process requires that the pool of instructions be constantly affected by very small random external perturbations so that chance adaptation to nondeterministic environmental changes can take place and prevail statistically.

4.1 Patterns, Images, Maps and Feature Detection

Before we continue with details about living systems, we must discuss some general properties and related definitions concerning information-driven interactions and the related dynamic concept of information. We said that in an information-based interaction, the presence of a given pattern in complex system A causes a specific change in complex system B. No direct energy transfer is involved, although energy must be supplied to the different stages of the interaction process (Fig. 3.7). The pattern in question can be any set of things, forms or features that bear and maintain a certain topological or numerical relationship in space and/or time (remember that we are *not* considering patterns made with a purpose by humans or programmed machines). We may represent a pattern in simplified form as a function of space and time $P_A = P_A(r, t)$. P represents a physical variable that can be simple (e.g., light wavelength or brightness, sound frequency or amplitude, etc.) or extremely complex (e.g., the spatio-temporal distribution of electrical impulses in brain tissue). In particular, the pattern may consist of a discrete spatial or temporal *sequence* of objects or events from a given, finite repertoire (e.g., the bases in the DNA molecule, notes of a melody, phonemes of a language or the action potentials fired by a neuron). In such a case, we may represent the pattern by a multidimensional vector $P_A\{s_1, s_2, \ldots, s_k\}$, where the s_k could be positions, instants of time, pitch values, neural impulse frequencies, etc. If $P_A(r, t)$ is a continuous function defined over a two-dimensional surface (e.g., the limiting surface of an object), we may call it a *symbol* or *sign*. Some purely spatial patterns do not require a local supply of energy to persist in stable form, whereas temporal patterns in general do. In general, when a pattern is the trigger of an information-driven interaction, we say that the corresponding pragmatic information is *encoded* in that pattern or *expressed* in it, or *represented* by it.

It is tempting to introduce the minimum number of bits needed to describe the function $P_A(r, t)$ as a measure of the algorithmic information *content* (Sect. 1.4) of pattern P_A. That, however, would only represent the information content of the pattern itself as an object, but not of the information actually represented by the interaction A → B. In other words, it would *not* represent any correspondence and it would be detached from the meaning we have given to the concept of pragmatic information in Sect. 3.5. For instance, the complex system A (the "sender") could exhibit many different patterns which, however, from the energetic point of view were all equivalent (*a priori* equiprobable) and had all the same algorithmic information content. However, changing from one pattern to another would change the response of B (the information) even if this would not represent any change in the macroscopic thermodynamic state (energy) and the information *content* of the system (we already encountered an example of this with the DNA molecule in Sect. 1.4).

Let us now consider a broad category of information-based interactions in which the change in B triggered by pattern P_A consists of the appearance of *another pattern* $P_B = P_B(r, t)$. We call P_B the *image* of P_A and say that in this interaction "B has received information from A, processed it and transformed it into a pattern P_B." A condition is that the correspondence between the two patterns be univocal: P_B is triggered by P_A and *only* by P_A (for multiple correspondences, ambiguous responses in neural systems and noise effects, see Sect. 4.3). If the pattern elicited in B is geometrically identical to that in A, we call the process a *replication* or *copy* (although we have promised not to deal with man-made things, a xerographic copy is an example). If it is only partially identical, the process is a *transcription* or *partial copy*. If we now use the output pattern P_B as an input to a similar interaction mechanism, we can obtain further identical copies: This process is called *reproduction*. In each case, errors may be introduced, which compromise the quality of the image and gradually propagate and increasingly obliterate the "offspring" in a reproduction. For the moment, we shall ignore the occurrence of such perturbations, although they do play a fundamental role in Darwinian evolution.

In the general case, when there is no point-to-point, instant-by-instant or feature-to-feature correspondence between both patterns (the uniqueness of the correspondence remaining, however), we call the process *mapping,* with the image, i.e., pattern P_B, a *map* of P_A. A hologram is a man-made example – it looks very different, but still maintains a univocal correspondence with the object, and a reverse transformation can restitute the original pattern). In genetics, this process is called a *translation* (careful: in physics the term translation has a specific geometrical/kinematic meaning that does not apply in this case). Finally, we may have a correspondence in which there is a response pattern P_B only for certain, well-defined input patterns $(P_A)_n$. In that case the system is a *feature detector,* the features being the special and unique patterns $(P_A)_n$ to which the system responds. In all this, what is being imaged, copied, mapped, reproduced or detected is information: It is what remains invariant throughout the related interaction processes, thus embodying the uniqueness of the correspondence (Fig. 3.7).

After this rather dry presentation of definitions, a few examples are in order. Instead of turning to the more basic molecular level, we shall begin with some highly oversimplified examples from the neural system, because we can refer to the more familiar examples of sensory experience. Consider the bundle of neural fibers shown running horizontally in Fig. 4.1a. This could be, for instance, a representation of the fibers of the acoustic nerve, originating in the spiral ganglion of the cochlea and carrying information on the vibration pattern along the resonating basilar membrane [94].[†] We assume, just for simplicity, that each fiber can carry only one type of standardized binary signal

---

[†] Most books and review articles on neural information use examples from the visual system. We also show features of the acoustic system, which encodes stimuli in *only one* spatial dimension: Each nerve in the peripheral acoustic system receives stimuli from a spatially limited region along the one-dimensional extension of the basilar membrane, responding to acoustical energy in a narrow region of the one-dimensional frequency space (see [94]). The optic nerve, on the other hand, receives signals distributed on the two-dimensional retina. In both cases, though, information is also encoded in the form of firing frequency; as a matter of fact, the acoustic system has extremely sophisticated temporal encoding mechanisms with resolution down to tens of microsecond, whereas the optic sense is rather dull in this respect (see Sect. 6.2).

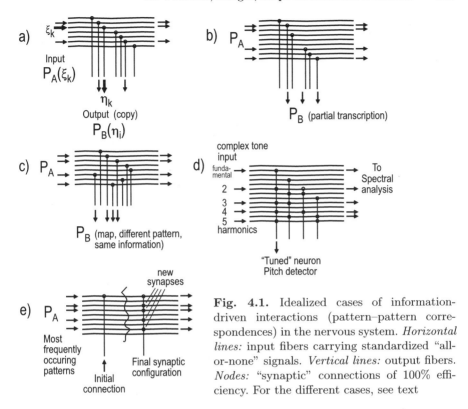

Fig. 4.1. Idealized cases of information-driven interactions (pattern–pattern correspondences) in the nervous system. *Horizontal lines:* input fibers carrying standardized "all-or-none" signals. *Vertical lines:* output fibers. *Nodes:* "synaptic" connections of 100% efficiency. For the different cases, see text

(on–off, or as neurobiologists say, "all-or-none"), ignoring for the time being other modes of encoding such as the particular distribution in time (firing sequence) or some analog variable (firing frequency), which we shall discuss in Sect. 4.2. The full input pattern across the bundle of fibers at a given instant of time could then be represented by a vector $P_A\{\xi_1, \xi_2, \ldots, \xi_k, \ldots \xi_n\}$, in which ξ_k is 0 or 1 depending on whether fiber k is firing an impulse or not, respectively (see next section for a more realistic representation). Now consider this input bundle connected synaptically to an output bundle in a very regular diagonal form as shown in Fig. 4.1a. It should be clear that the output distribution will be identical to that of the input: $P_B = P_A$; this is an example of a pattern copy. We can cross-wire the input fibers many times in the same way and obtain a pattern multiplication or reproduction system. In Fig. 4.1b we show the case of a system mediating a partial transcription. Note that in this case different input patterns can, under certain circumstances, lead to identical outputs.

A network connection like the one shown in Fig. 4.1c, which could represent a random wiring scheme, is the example of an interaction where the output distribution will be completely different from that of the input. However, since the correspondence is univocal, the information carried by the output pattern will still be the same. This somewhat trivial realization is of fundamental importance for a better understanding of brain function, for which we tend to have this "pseudo-intuitive" expectation that, for instance, a visual image is somehow preserved in its geometric form throughout the mental processing stages. But the neural image of a tree does not in any way preserve the topological features of a tree much beyond the eye's retina and first stages in the afferent pathways! On the other hand, even the most abstract thought is represented by very concrete patterns of electrical neural activity that stand in unique correspondence with the occurrence of that thought. And the neural activity distribution elicited in your brain whenever you see red might be totally different from the one elicited in my own brain under equal circumstances – yet for both of us it is perceived as red and nothing else! What counts are the *ultimate effects* of the corresponding neural patterns: the final output at the end-point of the intervening chain of information-driven interactions. We shall take up this discussion later and again in Chap. 6.

Our next example, the wiring scheme shown in Fig. 4.1d, is a little more realistic. It represents a *filter* or   system. Let us assume that to elicit a response in any one of the output fibers *all* its connections must be activated at once; as a consequence, each output fiber is "tuned" to a certain input pattern and will respond only if that particular pattern is present, remaining silent to all others. Feature detection is a fundamental operation in the visual and auditory neural systems. For instance, a cross-wiring in the cochlear nucleus of the acoustic pathway somewhat similar to the first vertical fiber in Fig. 4.1d is believed to play a role in the pitch detection of complex tones. These are periodic tones made up of a fundamental and a series of harmonics of frequencies that are integer multiples of the fundamental; the human voice, animal utterances and musical sounds are made up to a large extent of such complex tones. The superposition of harmonics sets the basilar membrane in resonant oscillation at different places at the same time, one for each lower harmonic; as a result, each complex tone is represented by a very specific and complex distribution of neural activity in the fibers of the acoustic nerve. Yet under normal listening conditions, we perceive only *one* pitch: that of the fundamental. Indeed, if a cross-fiber (Fig. 4.1d) is wired only to those input fibers which are activated by the harmonics of a given complex tone, that output fiber becomes a "detector" of that particular tone and would be responsible for the elicitation of a single pitch sensation.[†] It also will respond if not all harmonics are present, leading to a single pitch sensation for an

---

[†] Complex tone pitch processing is far more complicated than described above; there is a parallel mechanism of a time representation, based on a nonlinear response of the primary detectors (hair cells) on the basilar membrane, which works mainly in the low-frequency region (e.g., [94] and Sect. 6.2). However, the greatly simplified example of Fig. 4.1d is sufficient to illustrate some fundamental characteristics of a feature detector.

incomplete input in which, for instance, the fundamental could be absent – a fact amply verified in psychoacoustic experiments that demonstrate the self-correcting capacity of the acoustical information-processing system [94]. It is generally believed that this circuitry is not genetically prewired but *acquired through experience* by the systematic exposure to harmonic tones already beginning during intrauterine life (unassisted learning; see Sect. 4.3).

I want to emphasize in passing that here we have come across a concrete example of how information on the environment – in this case, the acoustic environment – is "built into" a living organism: A correspondence has been established between some very specific class of input patterns (the harmonic tones) and a particular architecture of the pertinent neural information-processing network. If an organism were to be brought up in a strange acoustic environment in which only *inharmonic* tones of a certain class would exist (for instance, sounds whose overtones have frequencies that are in a fixed but *non*integer relationship with the fundamental frequency), a different neural wiring scheme would emerge, with a different kind of pitch values and "musical" interrelationships. In other words, an adaptation to a different acoustic *niche* would take place. In our example, as stated above, the information is built into the organism as a late-fetal and early postnatal adaptive process; this is different from storing information during an assisted learning process, although both types involve synaptic changes, as we shall see below. In Sect. 4.4 we will discuss the more fundamental case of environmental information acquired very slowly during the evolution of a species. Many examples of evolutionary neural development can be found in the visual system; indeed, most optical networks up to the visual cortex are genetically prewired and accomplish a variety of tasks such as detection of motion, linear luminosity gradients, specific geometric forms, etc. [62, 72]. In those cases the information on the environment has been incorporated though the evolutionary process (we should note that neural networks for feature detection are much better known in the optical system because of their greater neuroanatomical accessibility to invasive experimentation with laboratory animals). We shall come back to this issue of "building information into the structure of an organism" in Sect. 4.4, when we discuss molecular information.

In the previous paragraph we mentioned something about neural hardware changes "acquired through experience" and "learning." A fundamental property of most information-based interaction systems in the central nervous system is the ability of changing in a specific, deterministic way as they are activated repeatedly with similar input patterns. The most idealized "act of unassisted learning" consists of the establishment of a connection between two originally independent neurons when for some externally controlled reason they tend to fire pulses simultaneously (the Hebb hypothesis [56]; a more realistic description will be given in the next section). For instance, in Fig. 4.1c, if the signal in fiber I_1, which does generate an output signal in fiber O_1, is frequently accompanied by input signals $I_2, I_3, \dots$, connections will appear in time at the nodes N_2 and N_3, with the end result that fiber O will be "tuned" to the entire input *pattern* $\{I_1, I_2, I_3, \dots\}$, and may be triggered in the future by only a partial input, say $\{I_2, I_3\}$. We may say that the system *has acquired a memory* of the frequently occurring complex input pattern $\{I_1, I_2, \dots\}$ and that it will respond correctly if presented with only a part of it (this is called autoassociative recall – see later).

The input–output relationships depicted in the simple configurations shown in Fig. 4.1 can be described mathematically with linear models, in which an input vector $\boldsymbol{I}(\xi_k)$ is related to an output vector $\boldsymbol{O}(\eta_i)$ through a matrix R_{ki}:

$$\eta_i = \sum_{k=1}^{N} R_{ik}\xi_k \,. \tag{4.1}$$

If the matrix elements (line interconnections) are given, they would represent a neural network with inborn (genetically programmed) wiring. If the matrix elements are amenable to change during usage, the network is adaptive, and two categories may be envisioned: 1. unassisted learning with changes of the matrix elements R_{ik} determined by the Hebb rule mentioned in the previous paragraph, i.e., changes governed by the frequency of usage of given interconnections; 2. changes controlled by some built-in information feedback from the output (also called "back-propagation," stepwise corrective changes in order to eventually achieve a desired output pattern for a given input). Sometimes the output is fed into a discriminator, to achieve further filtering (as we implied in the case of Fig. 4.1d, when we required that, to fire, an output fiber had to be excited simultaneously at all of its synaptic interconnections). Finally, we should clarify that although thus far we have only considered as input or output individual standardized impulses, relation (4.1) is mainly used with continuous variables, like neural firing frequency (see the next two sections).

4.2 Representation of Information in the Neural System

We have been getting ahead of ourselves. Without entering into any physiological details, it is time to describe a few generalities of neural system operation. To endow motile multicellular species with a faster and more complex information processing capability and enable memory storage of current events, a nervous system evolved together with a sensory and motor machinery to couple the organism to the outside world in real time. At the earliest stage of evolution the nervous system merely served as a "high-speed" information transmission device in which an environmental physical or chemical signal, converted into an electrical pulse in a sensory detector, is conveyed to a muscle fiber triggering a contraction. Later, neural networks evolved that could analyze and discriminate among different types of input signals and, depending on the case, send specific orders to different effectors. Finally, memory circuits emerged, that could store relevant information acquired during the organism's lifetime to successfully face previously experienced environmental challenges again at a later time.

The basic building block and information processing unit of a neural system is the *neuron,* a cell with an electro-chemically active membrane capable of generating short pulses of electric potentials that serve as information carriers. In standard texts on neural networks one usually introduces a "formal" or "ideal" neuron [73] – a cell *model* that bears some salient features of most (but not all) neurons, from jellyfish to human cortex. These main features are (Fig. 4.2): 1. a *dendritic tree,* collecting signals that arrive through synaptic connections with "upstream" neurons; 2. the cell body or *soma* which contains the nucleus of the cell and carries out protein synthesis and metabolic functions, and which also may receive direct input from presynaptic neurons; and 3. the *axon,* a process that conveys standard-sized electric signals away from the soma to postsynaptic neurons. The neuron is thus the fundamental information collection, integration and discrimination unit of the nervous system.

The fundamental biochemical process that makes this possible is the diffusion of metallic ions (K^+, Na^+, Ca^{++}) through the neural membrane. When the cell is "inactive," i.e., in a dynamic equilibrium, there is a metabolically maintained 35 times excess concentration of potassium ions and 10 times deficit of sodium ions in the cytoplasm with respect to the medium outside; K^+ ions will thus tend to diffuse out and Na^+ into the cell. But the permeability of the cell membrane for the potassium ions happens to be almost ten times higher than for sodium ions; since both travel through separate channels in the membrane, one can draw a simple equivalent d.c. circuit (in which the ion interchange process, or "sodium pump" is represented by an e.m.f. whose internal resistance is proportional to the inverse of the permeability) and calculate the *resting potential* V_r of the neuron, which is the electric potential difference between its cytoplasm and the outside (taken as the "ground potential"). For an average neuron it lies between $-70\,\text{mV}$ and $-90\,\text{mV}$; it is negative, because more K^+ flow out of the cell than Na^+ ions can come in per unit time. This constant flow requires a continuous metabolic supply of energy from the cell. The value of V_r is enormous, taking into account that the membrane is only about 70 nm thick (it represents an electric field across the membrane of $10^6\,\text{V/m}$ – no wonder I sometimes feel as if sparks were flying in my head!), but the potential decays rapidly with distance outside the membrane.

When a neurotransmitter is released into a synaptic gap, a local change in the ion permeability of the postsynaptic membrane causes a local change ΔV of the membrane potential, which will propagate away from the site as a "wave" of imbalance in the potassium/sodium concentration. Some synapses

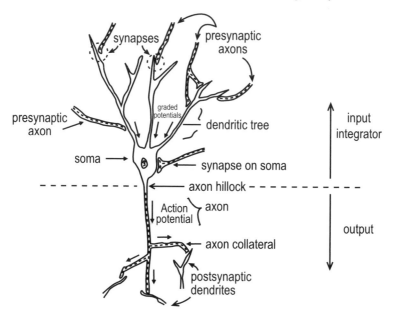

synapses

presynaptic
axons

presynaptic
axon

graded
potentials

dendritic tree

input
integrator

soma

synapse on soma

axon hillock

Action
potential

axon

output

axon collateral

postsynaptic
dendrites

Fig. 4.2. Sketch of a "formal" neuron

(in which the neurotransmitter is dopamine, glycine or gamma-aminobutyric acid) produce a positive pulse ΔV, called *excitatory postsynaptic potential* (EPSP), leading to depolarization (making the cell's interior less negative); other synapses (with a neurotransmitter like norepinephrine) produce negative ΔV pulses or *inhibitory postsynaptic potentials* (IPSP), leading to hyperpolarization. EPSPs are of the order of $+8\,\mathrm{mV}$ to $+10\,\mathrm{mV}$, IPSPs $-5\,\mathrm{mV}$ to $-8\,\mathrm{mV}$; they reach a peak in a few milliseconds and decay within about $10\,\mathrm{ms}$. The *efficiency* (or potency, efficacy) of a synapse (relative amplitude of the generated signal) depends on the actual structure and size of the synaptic junction and on the number of neurotransmitter receptors on the postsynaptic cell membrane. As a matter of fact, the actual amplitude and duration of a postsynaptic potential is a measure of the potency of the synapse where it has originated, which, as mentioned above, can change during a learning (or forgetting) experience.

The incoming pulses add up more or less linearly as they travel to the cell body.[†] However, there is a limit to this linear behavior in all excitable membranes: If because of some perturbation the cell potential is depolarized above a certain threshold value (this can be done artificially with a microelectrode), the permeability for Na^+ ions increases suddenly, and an avalanche

---

[†] Careful with this image of "traveling" impulses: Their duration is usually longer than the time it takes them to propagate to the soma; therefore, the picture is closer to that of a neuron "lightening up" as a whole!

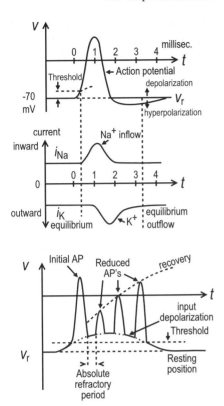

Fig. 4.3. *Upper and middle graphs:* sketch of the generation of an action potential, the standardized output pulse of a neuron and "unit of communication" in the nervous system (cross-membrane voltage and ion current, respectively). V_r is the resting potential of the cell's cytoplasm. *Lower graph:* sketch of the generation of a volley of APs by a lasting input depolarization. Although molecular processes are involved, the mechanism is purely classical

of sodium into the cell occurs, creating a large positive depolarization pulse. This threshold is believed to be smallest at the axon hillock, and it is precisely there where the positive *action potentials* (AP) are generated. The triggered instability is independent of the triggering event, and runs its course down the axon in standardized digital ("all or none") fashion.[†] Figure 4.3 sketches the time-dependence of an AP at one given point of the axon membrane. As can be seen, the Na^+ inflow is rapidly compensated by an increased K^+ outflow, until the resting state is again achieved. If the original (EPSP) depolarization lasts beyond the duration of an AP, a volley of output pulses can be launched, but of lesser amplitude (lower graph in Fig. 4.3). The speed of propagation of an AP ranges from about 0.3 m/s in the "bare" axons of cortical tissue to about 130 m/s in the myelinated fibers of efferent (outgoing) nerves. There is evidence that the AP-generating process can also launch pulses into the dendritic tree, i.e., generate retropropagating impulses; this may be relevant to the learning process (Sect. 4.3).

[†] Careful again: The AP does not represent a transport of energy. It is only a propagating signal; the energy for the required ion exchange is provided locally by the neuron's "sodium pump" as the signal passes by.

Electrical impulses are not only generated in neurons. There is another fundamental type of "pseudo-neurons" with excitable membranes, the *receptor cells*, in which electric potentials are triggered by some *physical* agent, like a photon in the case of rod and cone cells of the retina; a flexing force on the cilia of the hair cells of the basilar membrane or on the stretch receptors in muscle spindles; or by an external *chemical* agent in the case of olfactory receptors in the nose and blood composition monitors in the hypothalamus. These neurons are the sensory and proprioceptive *detectors* of the nervous system and represent the input extreme of the neuronal chain of an animal, feeding information on environment and body through the peripheral afferent pathway system to the information-processing machinery of the central nervous system. On the output extreme of this chain are the special synapses (with acetylcholine as transmitter substance) between axons of so-called motoneurons and the muscular fibers in muscles, glands and blood vessels; these synapses, when activated, generate the contraction of motile proteins, collectively causing the contraction of these fibers. Motoneurons are neural system *effecters* that control the actions of the organism *on* the environment – in other words, they control the behavioral output.

A formal neuron has been viewed as a logic gate with a standard digital output triggered when the linear summation of the cell's inputs surpasses some threshold value. In reality, even model neurons require a more accurate description and must be treated as complex *analog* devices. There are several reasons. First and foremost, the fundamental output information of a neuron is not encoded just in the form of individual AP signals, but rather in their *temporal* sequence or time distribution. The relevant variable in question could be the average firing rate (with the information actually expressed in the form of changes in that rate, or frequency *modulation*); for instance, in certain circuits of the auditory system, the information is encoded in the distribution of periodic neural spike volleys in synchrony with the fundamental frequency of the tone perceived (e.g., [99]). In the literature there is often mention of the noise in the neural communications environment. But one has to be careful: 1. noise itself (say, stochastic departure from periodicity) can be a deliberate carrier of information; 2. neural networks are robust in the sense that they are efficient noise suppressors (see later, e.g., Fig. 4.7). Many neurons also have a *spontaneous* firing rate in absence of any input, which is a characteristic parameter of their own. Inhibition or decrease of this spontaneous rate is as valid an expression of information as is rate increase. Another requirement for a model neuron is related to the fact that in a learning act the synapses are not just established and remain 100% efficient thereafter, as for instance it was assumed in the example of Fig. 4.1e, but that their efficiency or potency may change rather gradually in either direction during a learning process. As anticipated in the previous section, this neural or synaptic "plasticity" is the basis for long-term memory. In other words, not just the architecture of the synapses may change, but the efficacy of the individual synapses as well. This has been summarized in the mathematical expression of one neuron's ideal "transfer function" (e.g., [63]):

$$\eta = \eta_s + \sum_{i=1}^{i} \mu_i \xi_i \qquad (4.2)$$

in which η is the frequency of the output APs, η_s the spontaneous firing rate in absence of any presynaptic input, ξ_i is the input frequency received at synapse i, n is the number of these synapses and μ_i the synapse's efficiency (> 0 for EPSP, < 0 for IPSP), which because of plasticity may be a function of time (see relation (4.3) below). It is necessary to modify accordingly relation (4.1) for each node of the examples in Fig. 4.1.

There are three spatial domains of information-processing in the nervous system. The neuron represents the fundamental processing unit in the "microscopic domain."[†] At the "mesoscopic level" we have assemblies of relatively few neurons wired to each other to accomplish a limited repertoire of specific tasks. For instance, in the retinal network we find groups of a few neurons working as motion detectors or contrast enhancers; in the sensory receiving areas of the cortex there are columnar assemblies of neurons which represent cooperative units receiving information from a common, limited stimulus region of the sensory detector (retina, basilar membrane, skin, etc.), which may comprise hundreds of thousands of neurons. The "macroscopic domain" is the brain as a whole, which in a human has about 10^{11} neurons with 10^{13} to 10^{14} synaptic interconnections. This progression of domains is not dissimilar (except in absolute scale) to the one we find in the material structure of macroscopic bodies: the microscopic domain of individual atoms or molecules, the mesoscopic domain with small-scale fluctuations, and the macroscopic domain, accessible to our senses (Chap. 5). But in fundamental contrast to the physical case where there may be "zillions and zillions" equivalent states of microscopic configuration compatible with just one macroscopic state (Sect. 5.4), in the neural system, each microstate – the spatio-temporal distribution of neural activity – has a unique meaning and purpose! Still, as we shall see in Chap. 6, there is something "macroscopic" that does bind them together informationally and in their temporal succession: animal consciousness; and there is something in one particular species that controls them collectively: human self-consciousness.

There are two basic ways in which information is represented in the nervous system at the meso- and macroscopic levels. Type 1 is a dynamic form given by the specific *spatio-temporal distribution of electrical impulses* in the neural network, representing the *transient* or *operating* state of the neural network. Type 2 is represented by the *spatial distribution* and the *efficacies* of inter-neuron connections in the neural tissue (the synaptic architecture),

---

[†] What is microscopic here is still at least 10^5 times bigger than what would usually be called microscopic in physics.

representing the *internal* state or *hardware* of the neural network. Type 1 is a pattern that varies on a time-scale of a few to several hundreds of milliseconds and usually involves millions of neurons even for the simplest information-processing tasks, requiring a large and continuous supply of energy to be maintained (it is the increased vascular blood flow and the oxygen consumption that appear mapped as images in functional magnetic resonance (fMRI) and positron emission tomography (PET), respectively, [57, 81]). Type 2 is a slowly varying spatial pattern and may require a much smaller supply of energy. As we shall see later, especially in Chap. 6, there is no equivalent to "software" in the neural system – the "programs" are embedded in the changing configuration of the neural hardware (synaptic architecture).

Unfortunately, none of the macroscopic neural patterns, dynamic or static, can be expressed adequately in a mathematical form. This task is hopeless for various reasons. First, type 1 could not be approximated by any discrete or continuous distribution function because neighboring neurons, while usually encoding similar information, may do so at very different firing rates or impulse distributions. Second, in the human brain there are over 10^{10} neurons in the cortex, each one connected to thousands of others, which means that there could well be 10^{13} to 10^{14} synaptic connections at any given time to be described in terms of their position and efficiencies. Third, there are also insurmountable difficulties from the experimental point of view. A single microelectrode implanted in a neuron (which does not necessarily impair the function of the cell) will give us readings of individual neural pulses, their frequency and distribution in time, but it would take thousands of microelectrodes implanted in very close vicinity of each other to obtain a real "spatio-temporal distribution" of impulses in one processing module of brain tissue. Larger-tip electrodes will provide measurements of electrical signals averaged over tens or hundreds of neurons and will tell us what the average activity of a limited region of neural tissue or nerve fiber is at any given time – but it still would not furnish any details of the full spatio-temporal distribution of impulses. Concerning the neural architecture that represents the type 2 information, synapses can only be observed with a microscope, and although synaptic growth has actually been studied in real time [60], only statistical results about synaptic architecture have been obtained, and this in only very limited parts of neural tissue. We should note that defining average quantities like densities for the mathematical description of the "state of the brain" does not seem to make much sense, except that the signals responding to a vast number of individual sources, such as the evoked potentials (EEG) and the tomographic images (fMRI or PET), are of fundamental importance to clinical diagnostics and the overall understanding of brain function.[†]

[†] The noninvasive techniques with external electrodes in electroencephalography, or the small SQUID magnetometers in magnetoencephalography, while only providing information of activity averaged over centimeter-wide areas of the brain, have the advantage of being fast, and thus can be used conveniently for timing

All of the above mentioned difficulties, however, shall not discourage us from using the concepts of spatio-temporal distribution of neural activity – which we shall henceforth call "neural activity" – and the spatial distribution of synapses and their efficiencies – the synaptic architecture – as the two fundamental expressions of information in the nervous system. But there must be a clear understanding that at present any mathematical representation thereof can only be a very crude approximation of physiological reality. Still, this will not prevent me from trying to convince the reader in Chap. 6 that the dynamic spatio-temporal distribution of neural activity and the quasistatic spatial distribution of synapses and their efficiencies together are the physical realization of the *global state of the functioning brain* at any instant t – which embodies all mental acts and is what makes you *you*. No extra players with indefinable roles are needed (such as "mind" and "soul") ...

In general, even in the simplest nervous systems of invertebrates, there are several mutually interacting levels at which information is represented or mapped. Each level serves well-defined purposes and performs related tasks, each one transforms certain well-defined input patterns into well-defined output patterns (thus representing the information-processing operations per se), and in many instances, the synaptic distribution actually handling the pattern transformations changes with time deterministically as a function of the actual informational tasks performed (adaptation or learning process). A significant characteristic is the existence of *feedback* channels at all levels, which play a fundamental role in the control of synaptic plasticity and in the memory recall process, as we shall discuss in the next section.

The basic processing unit is the neuron, as described above. The highest processing level in the mammalian brain is the cerebral cortex. The anatomy (interneuronal wiring scheme) of the cortex is adapted to its operations; it is described in detail in many textbooks (for instance, [23, 62]). Let us just mention some significant properties:

1. most of the neural input from distant regions of the brain is delivered at the top of a cortical column and in each column, information is largely processed vertically down;
2. through short intracortical axon collaterals, pyramidal cells form networks that operate as a group;
3. through cortico–cortical fibers, one cell may contact several thousands of other cells in distant regions of the cortex;

---

estimates of the order of milliseconds. fMRI and PET records elicited by a very brief, localized neural activity pulse (millisecond) reach their maximum after 5 s and decay within about 12 s (e.g., [57, 81]). The synchronism of neural firing rate over large volumes of neural tissue evidenced in EEG recordings of evoked potentials in various frequency ranges (alpha waves from 8 Hz to 12 Hz; beta waves from 15 Hz to 25 Hz and gamma waves from 30 Hz to 70 Hz) could play an important "global" informational role (e.g., [62]; see also Sect. 6.1).

4. some axons branch upwards in a column, providing upstream feedback information;
5. there are other types of cells like interneurons that have a specifically excitatory or inhibitory function;
6. the all-important glial cells have no direct involvement in information-processing but feed nutrients to all neurons.

Finally, there are the incoming and outgoing projections, fiber bundles carrying input and output signals, respectively, connected to distant regions of the brain and organism.

4.3 Memory, Learning and Associative Recall

In this section we shall concentrate on some very basic integral information-processing operations of the neural system: environmental representation, memory storage and recall. These operations are so fundamentally different from the equivalent ones with which we are familiar (say, photography, computer memory storage and document downloading, respectively) that it takes a novice some time to develop an intuitive understanding of what are called the "holographic mapping,"[†] "distributed memory" and "context-dependent recall" modes of the nervous system. Let us assume that we have an input (sensory) system S (Fig. 4.4) in which physical patterns of sensory stimulation are converted into homologous patterns of neural activation. This for instance could be the retina of the eye or the organ of Corti (hair cells) on the basilar membrane in the cochlea. This system interacts with a number of consecutive stages of neural information processing which we designate as systems A, B, C, ... We assume that the neural patterns in the first stage S are exclusively of transient type 1 above (spatio-temporal distribution of neural signals) whereas the patterns at the higher levels A, B, C, ... may also include those of type 2 (changes in the synaptic architecture during their use). Although we have said that it is impossible to express a neural pattern as a mathematical function that is true to physiological reality, let us adopt a simplified model and designate with η_A^k the firing frequency of an outgoing neural fiber k at stage A. When a pattern $P_A\{\eta_A^k\}$ is activated in A by a pattern $P_S\{\eta_S^k\}$ in the input channel (such as the physical signal representing the image on the retina of a visually sensed object, or the oscillation pattern of the basilar membrane in response to a musical tone), a specific pattern P_B is triggered in B, P_C in C, and so on. Clearly, we are in presence of a chain of information-driven interactions (Sect. 3.5).

If these patterns persist for an interval of time considerably longer (say, a few seconds) than the actual duration of the input pattern P_S we say that P_A, P_B, P_C, ... are *short-term, activity-dependent* or *working memory*

[†] To avoid confusion with optical holography (see Fig. 4.8), one should really use the word *holol021ogic*.

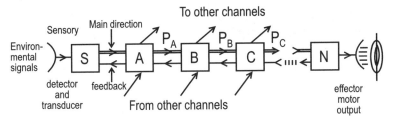

Fig. 4.4. Hypothetical set of sensory information processing stages in a nervous system. S is a sensor in which a physical signal (photon flux, acoustical energy, muscle spindle stretch, etc.) is converted (transduced) into neural impulses. A, B, C, . . . are sequential stages, where one neural activity pattern P is converted into another. In most brain networks there is also a feedback system (and input from other channels). If neural activity patterns specific to a given sensory stimulus persist for several seconds after the original stimulus has disappeared, we say that the corresponding information is stored in short-term or "working" memory. If stimulus-specific patterns can be triggered much later by activity in some "upstream" stages, we say that information on that particular stimulus was stored in long-term memory; reelicitation of the corresponding patterns represents the recall process. Somewhere at the other end of the chain there may be connections to muscle fibers and glands whose activation would represent part of a behavioral output

images of P_S. We further assume that, as already mentioned above and verified for most central nervous systems, there is an information-based feedback relationship between the consecutive neural levels A, B, C, such that when, for instance, pattern P_B in B is evoked in some independent way at a later time, the original pattern P_A to which it corresponds, is elicited, at least in part, in A. If this is the case, we say that information on P_A, and hence also on the corresponding sensory input P_S, was stored in *long term* or *structural memory;* this long-term image would have to be of type 2, i.e., a change in the actual hardware of synaptic connections (whether it is establishment of new synapses or changes in the efficacy of existing synapses will depend on circumstances; e.g., rats raised in challenging environments have a higher spine count (related to density of synapses) than animals raised in "boring" environments [48]). The feedback process, in which the original patterns P_B, P_A, etc. are being reconstructed *without* the corresponding full external input P_S, is called a *memory recall.* In other words, the memory recall of a sensory (or any other type) event consists of the *reenactment* of neural activity patterns that were present when that event was actually perceived. Again, this is a very different type of retrieval than any of the processes with which we are familiar in daily life. To retrieve the photograph of my grandmother, I must know its "address" in the family album, go there and access the image physically. To mentally recall or remember an image of the face of my grandmother, my brain must recreate at least part of the neural activity that was in one-to-one correspondence with that elicited by the actual perception of her face (which, as mentioned before, bears no topographical similitude (retinotopical mapping) with the features of the photographic image). What is stored in the brain is not the perception-triggered pattern itself, but the *capacity to regenerate* that pattern. There also could be a process stretched in time: In the so-called *procedural memory* recall, a given motor skill is executed during an extended period of time (riding a bike, playing piano without paying attention to the music played, driving a car while talking on a cell phone, etc.). The capacity to regenerate a neural pattern has been called the "principle of virtual images" (e.g., [62]); it is also the basis for the explanation of optical and acoustic illusions (e.g., [37]). In my view, it is no longer a principle but an experimentally proven fact – and the images are not "virtual" but real (see Chap. 6). Of course, there is no way to demonstrate that exactly the same complex spatio-temporal activity distribution is recreated every time an image is recalled, yet single cell recordings (and to a certain extent, functional MRI and PET tomography) convincingly show that the same cell *clusters* that had been involved in the corresponding perception become active again once that specific image is being recalled by association, voluntarily imagined or just only expected.

A very fundamental neural processing operation is the formation of *associative memory,* in which the brain establishes the memory of a lasting correlation between simultaneous or near-simultaneous events, and to which other types of memory forms can be related. Suppose that there are two nearly simultaneous input patterns P_S and Q_S (e.g., the visual image of an object and the acoustic pattern of its name, respectively), triggering the different patterns P_A and Q_A, respectively, at the primary neural level A; P_B and Q_B at the secondary level B, and so forth. We now assume that when these input patterns are repeated several times in near-simultaneity, changes occur in the synaptic architecture at level B such that (see sketch in Fig. 4.5 [96]): 1. a *new* pattern $(PQ)_B$ emerges that is a representation of the simultaneously occurring stimulus *pair* P and Q; 2. this new pattern appears even if either *only* P or *only* Q is presented as input; 3. whenever pattern $(PQ)_B$ is independently evoked, *both* P_A and P_B will appear at the lower level A due to feedback. As the result, a sensory input P_S will trigger an additional response Q_A at level A even *in the absence* of input Q_S, and an input Q_S will trigger P_A (as feedback from level B). This is the essence of what is called an *associative recall.* The formation of a new image of the pair $(PQ)_B$ at the secondary level B is the result of a neural *learning process;* the change in hardware that made this possible represents the *long term storage* of information on the correlated inputs P and Q.

If a short-term memory mechanism is operating, P_S and P_Q need not be simultaneous – the new combined pattern can establish itself on the basis of the respective temporarily stored patterns (conditioning). Notice that P_S and Q_S could be inputs from two different sensory modalities as mentioned above; or they could be two different parts of the same object (e.g., the face

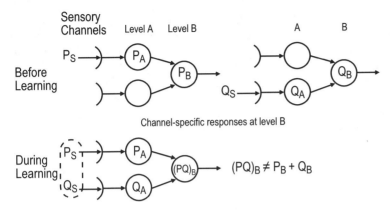

Channel-specific responses at level B

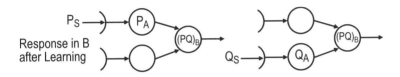

Repeated simultaneous input P_S + Q_S gradually changes hardware in B; a new response pattern $(PQ)_B$ to pair of stimuli emerges ("unsupervised" learning).

Elicitation of common response pattern $(PQ)_B$ by either sensory P_S or Q_S.

Fig. 4.5. Sketch of the basic mechanism of learning and associative memory in the neural system (from *Roederer* [96]). The change in "hardware" necessary to elicit the new, combined pattern (PQ_B) consists of a change in the spatial configuration and efficacy of synapses (neural "plasticity"). This new spatial configuration embodies the representation of information stored in long-term memory

of your dog and the body of your dog). One of the two inputs could be "very simple" compared to the other; in that case we call it a *key;* a very simple key as input can thus trigger the recall of a very complex image consisting of the superposition of different component patterns (e.g., a four-letter word, and the complex action it represents). This means that the replay of a specific neural pattern P_A can be triggered by cues other than the full reenactment of the original sensory input P_S – a *partial* reenactment may suffice to release the full image P_A (this is called *autoassociative recall*). This also goes for a partial input from the higher stages, and represents a fundamental property of the mechanism of associative memory recall (see Chap. 6). Finally, a very noisy image or part of image as input can trigger the recall of the "clean" original (see Fig. 4.7 below). These operations represent most basic algorithms for "intelligent information processing." They have been discussed extensively in the literature using numerical simulations with mathematical models of neural networks (e.g., [3, 63]).

Let us just describe a few aspects of the learning process, focusing on one of the stages depicted in Fig. 4.5 and remembering what we said in connection with Fig. 4.1e. There we defined as an idealized "act of unassisted learning" the establishment of a synaptic connection between two neurons that frequently fire at the same time [56] of a synapse i (see relation (4.2)) is proportional to the correlation coefficient between the pulses arriving at that synapse and the (independently triggered) firings of the postsynaptic cell. For a Poissonian distribution of impulses, this means that the time derivative of the efficacy of input synapse i is proportional to the product of input firing frequencies ξ_i and the axon output frequency η:

$$\frac{d\mu_i}{dt} = \alpha_i \xi_i \eta \qquad (4.3a)$$

α_i is the "plasticity coefficient" of the synapse in question. A somewhat more realistic model introduces the average background input rate ξ_b and sets the efficiency rate of change proportional to the difference $\xi_i - \xi_b$ [63], which can be positive or negative (a synapse that persistently "underperforms" would eventually be wiped out); the plasticity coefficient is assumed to be the same for the entire cell:

$$\frac{d\mu_i}{dt} = \alpha (\xi_i - \xi_b) \eta . \qquad (4.3b)$$

This is an "automatic" change that ultimately depends on the occurrence and configuration of input patterns. Of course, a condition is that the values of the efficiencies reach some finite asymptotic value rather than diverging or tending to zero everywhere. In other adaptive systems, such as "teacher assisted" networks, the changes of efficiency μ_i are controlled by feedback from the output stage(s) in such a way that, after the learning experience, the output for a given input pattern will have the desired form. Part of the feedback fibers in the cortex may play such a role; but by and large this "back-propagation" of information in a learning process, already mentioned at the end of Sect. 4.1, remains experimentally elusive. Some activated neurons have been found to release a substance (nitrous oxide) that can diffuse to neighboring neurons and could strengthen those synapses that are receiving impulses at the same time [106]; on the other hand, AP "echoes," (small feedback impulses fired upstream into the dendritic tree when an AP is triggered), may also play a role [108].

Let me try to summarize comprehensively the functions of an ideal macroscopic information-processing unit in the central nervous system. Consider Fig. 4.6, a simplified model for one of the processing stages of Fig. 4.4. In formal analogy with Fig. 3.5 (in which the billiard balls have been replaced with model neurons, and their mechanical interaction by the transmission of electrical impulses), we have an input layer of neurons (the "detectors"), an output layer (the "effecters"), and a complex system of interconnected

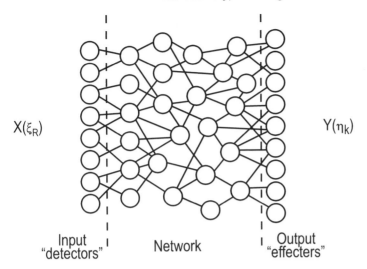

$X(\xi_R)$ $Y(\eta_k)$

Input   Network   Output
"detectors"   "effecters"

Fig. 4.6. Simplified model for one of the processing stages of Fig. 4.4 (compare with Fig. 3.5!), consisting of an input layer of neurons ("detectors" of neural patterns X), an output layer ("effecters" exhibiting corresponding patterns Y), and a complex system of interconnected neurons between them (the information processing mechanism). The "wiring scheme" may be genetically inherited, or it may be adaptive, i.e., able to change during a learning process. Linear networks like these are models for neural computing programs

neurons between the two. This represents an information-based interaction mechanism (Sect. 3.5), which establishes a univocal correspondence between an input pattern $X\{\xi_k\}$ and the elicitation of an output pattern $Y\{\eta_l\}$. ξ_k and η_l are the firing rates of input and output layer neurons, respectively. The actual relationship depends, of course, on the particular interconnection scheme and related synaptic efficiencies (just as the input–output relationship for Fig. 3.5 depended on the initial positions of the balls inside the "black box"). If the interconnections are "prewired," i.e., their design is inherited, the system evolved in a Darwinian long-term trial-and-error process to make the neural circuit accomplish some specific task beneficial for the organism (most feature detection operations, instinctive responses and the control of organ functions are governed by prewired neural circuits). We say that the internal structure of this piece of neural machinery "carries information" on the environment and the particular function it is meant to carry out. The information in question is a reflection of the most frequently occurring configurations and events in the organism's environmental niche.

If the neural network is plastic, i.e., its synapses change with use, information can be stored in the network during the lifetime of the organism through a learning process. As mentioned before, there are networks which can be conditioned to respond with specific, often simple, patterns Y to certain complex input patterns X that occur most frequently among all others. The complex-tone pitch processor in the auditory system is an example (Sect. 4.1), in which, after an "unassisted learning" experience (exposure to frequently-occurring special patterns), each input from a complex basilar membrane excitation elicited by the simultaneous harmonics of a complex tone, regardless of their individual intensities (the tone spectrum), will lead to a specific output pattern giving a single pitch sensation (equal to that elicited by the fundamental alone). A model of how this could be accomplished in neural tissue is described in [94]. Likewise, our visual system is able to recognize simple geometric forms (triangles, squares, crosses, etc.) as "one and the same thing," regardless of the size, orientation and position of their images on the retina. Particularly important in human development is the early capability of an infant to recognize the face of the mother as viewed from different angles. A reverse example, where a simple input pattern X triggers very complex images Y is the processing of human language.

Neural computation is a relatively new branch of computer science that develops programs for digital computers to emulate some fundamental operations of natural neural networks, especially pattern recognition. In these computer programs, a back-propagation process is built in through appropriate programming procedures which, however, bear little resemblance with what potentially may be happening in a live neural system. It is not our intention to describe and discuss the algorithms involved (e.g., see [58]), except to remark that much can be learned from the properties and operations of neural computation programs even if the mathematical models employed are highly simplified or are very distant forms of reality. One significant general result is that adaptive networks, even the most simple ones, have three universal intrinsic properties, which are also found in the natural neural circuits of the brain:

1. They are quite robust with respect to localized internal damage: The loss of information is not necessarily fatal, as it often happens with addressed memory systems (think of the photograph of a person whose face has been obliterated).
2. They are less sensitive to a noisy input (e.g., autoassociative recall with a noisy key may still work correctly, whereas a slightly wrong address number may be fatal).
3. They sometimes lead quite naturally to ambiguous or multiple output for identical input patterns (think of the "Necker cube" effect and other "switching" optical illusions, or some well-known ambiguous pitch perception effects, which have puzzled psychologists for a long time). Figure 4.7 shows examples of noise suppression (recall with a noisy key) and associative recall with a "damaged" key using computer simulation.

Finally, for the sake of completeness and to promote a better understanding of the all-important process of associative recall, it is instructive to discuss a physical model that has been proposed as an equivalent process (care-

Key   Recollection   Key   Recollection

(a)   (b)

Fig. 4.7. Examples of (**a**) noise suppression (recall with a noisy key), (**b**) autoassociative recall (with a "damaged" key), using computer simulation; from *Kohonen* [63]. The original images (3024 pixel each) were fed repeatedly into a linear neural learning network with a coefficient adjustment protocol until the condition output = input was achieved. With permission of the author

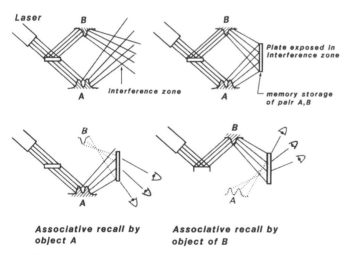

Associative recall by object A

Associative recall by object of B

Fig. 4.8. Scheme of a Fresnel holography setup (after *Roederer* [93]). *Top:* Two coherent laser beams illuminate objects A and B; a photographic plate is exposed to the intersecting beams. *Bottom:* Setup for image retrieval – when only one of the objects is illuminated with one beam, a virtual image of the other object can be seen. This optical process is conceptually equivalent to neural information retrieval by associative recall; for instance, B could be the visual pattern of your mother's face, A could be her name (the "key"). See also Fig. 4.5

ful: only *conceptually* equivalent!) (e.g., [87]). It is Fresnel holography (e.g., p. 273 of [63]). In Fig. 4.8 a split coherent laser beam illuminates two objects A and B (top of figure). A photographic image of the dispersed beams is made on a glass plate placed in the region of superposition. If E is the complex electric field magnitude at the plate (e.g., the electric field component in the direction of linear polarization), the relative change in transparency δT of the plate (a negative quantity) will be proportional to the amplitude $(E_A + E_B)(E_A^* + E_B^*)$ (starred quantities are complex conjugates – see Sect. 2.4). We assume that only the *phase* of E varies with the position r on the glass plate; in other words, $E_A E_A^* = E_B E_B^* = \text{const}$, independent of r. Under these circumstances we obtain $\delta T = -\lambda(2 + E_A E_B^* + E_B E_A^*)$; λ is a positive factor which depends on exposure time, etc. Now consider the lower part of the figure. We have removed object B, illuminate object A with the original laser beam, and look through the photographic plate in the direction of the former object B. Lo and behold, we will see a virtual image of the latter: The presence of object A has *recreated* the image of B! Any other object in position A, or the object A (which is playing the role of a key) displaced from its original position, will not recall the image of B! The same happens if we reverse the procedure: In absence of object A its image can be recreated by the presence of object B as a key for the recall! The physical explanation is as follows: $E_{\text{Aonly}} \propto (1 + \delta T)E_A = (1 - 2\lambda - \lambda E_A E_B^*)E_A - \lambda E_B$, which shows that the field corresponding to object B appears explicitly in the form $-\lambda E_B$, indicating that the information on object B has been retrieved. The image of object A appears with attenuated intensity (by a factor $(1-2\lambda)$), plus some noise from the interference term $\lambda E_A E_B^*$. The interesting fact is that it is possible to store several pairs of images (e.g., image + key) on *one* plate, and retrieve each one separately with the associated partner object or key; increasing noise, however will be creeping in. Something very similar happens in linear arrays of associative recall networks like those used for Fig. 4.7 – and in the real neural networks of the brain! Some neurobiologists have been overly enthusiastic with holographic models of the neural machinery, applying mathematical transformations common in optical holography (e.g., the Gabor transform) to brain operations [87]. This is unrealistic and unnecessary: Linear associative networks working on the basis of relationships like (4.1) to (4.3) with variable matrix elements and feedback all show the features that one finds in holography without having to appeal to lasers and other optical artifacts which have little physical equivalence in brain structures [63]. However, in defense of the "brain holographers," I must point out the pedagogical value of *any kind* of quantitative model – after all, as we shall discuss in the next chapter, all of science, including physics, is nothing but... model-making!

4.4 Representation of Information in Biomolecular Systems

It is high time to descend into the molecular realm – the other domain in which information-based interactions have appeared naturally and indeed have done so first during the evolution of life. Let us recall that in the neural system the correspondence established in information-driven interactions is of the type pattern → pattern (Sect. 4.1). The input or "sender's" pattern consists of a dynamic distribution of neural impulses in a given group of neurons or their axons. The transient pattern elicited by the interaction in the "receiver" also consists of a specific distribution of neural impulses in univocal correspondence with the input pattern. In addition, in the so-called adaptive or "plastic" neural networks, another more permanent type of pattern can be produced (Sect. 4.1), consisting of changes in the distribution and efficacy of synaptic connections between the participating neurons and representing the long-term storage of information (long-term memory). Exceptions can happen either at the beginning of a neural chain (Fig. 4.4), where the input pattern could be a physical signal from the environment, as well as at the end of the chain, where the output could be a pattern of mechanical contraction of muscle fibers driven by neural impulse-activated motile proteins in these fibers. Another possibility of output at this end is a neural impulse-triggered secretion of hormones – chemical messengers that use the bloodstream for the dissemination of information (see end of this section). The control of behavioral response and body functions is the main short-term purpose of the nervous system. A longer-term goal is the storage of sensory information for use at a later time to assure the possibility of an appropriate behavioral response during sudden and unforeseen environmental change. In humans, the initiation of an information-processing chain can be triggered from "inside" the system rather than by signals from the environment or the organism (the human brain goes "off-line" [12] during the human thinking process); furthermore, in humans the final product of an information-processing operation may not be intended, or even be usable, for motor output (e.g., abstract thinking, Chap. 6).

In the information-driven interactions of the biomolecular domain the input "pattern" is some static spatial order of chemical units or monomers in a macromolecule (often referred to as the "template") and the output usually consists of an action involving the assembly of another macromolecule. This latter molecule in turn could be the carrier of another pattern, as happens with a copy (DNA self-reproduction) or partial transcription (RNA transcription), or it could be the executor of a chemical/physical function (the proteins). The correspondence between input pattern and output action – in other words, the processing of biomolecular information – is in itself also mediated by macromolecules, which carry necessary *instructions* for their operation. The manufactured macromolecules must be assembled with subunits found in an appropriate medium (a cell's nucleus or cytoplasm). In essence, here we can trace an equivalence with neural networks: The information stored in the architecture of synapses and their efficacies embodies the instruction to transform a given input pattern into a certain output (this can even be applied to our billiard ball example of Fig. 3.5: The reset mechanism in the black box embodies instructions that link in consistent way a pattern at the detector with an elicited pattern at the effecter). In summary, from the informational point of view, in the microscopic domain we have molecules that are biological information carriers and molecules that are function carriers (information-processors and manufacturers of other molecules).

For information-carrier macromolecules of the same class (e.g., all DNA molecules) it is crucial that they be indistinguishable from each other *except* for the spatial order in which the molecular units that embody the information are arranged. This means that chemistry has to be indifferent to the particular sequence involved and that information-carriers of the same class all have to be energetically equivalent, i.e., represent, in relative terms, "pure information" (if they were to be assembled by chance – which of course they are not! – they all would be equiprobable, of equal Shannon information content – see Sect. 1.3). If this were not the case, interactions with other macromolecules would only be "physical," not information-based (see definition in Sect. 3.5). One condition sine qua non for these molecules is that the chemical units whose order represents the information must be *noninteractive* with each other as long as they are an integral part of the information molecule. There are additional structural conditions. One of them is that, for the information to be "read out" (i.e., copied or translated), the molecule must be a *linear* strand of units; bi- or tri-dimensional arrangements of the code would present extraordinary complications to any interaction mechanism. All this imposes enormous restrictions on the composition and structure of the macromolecules involved – in essence, only nucleic acids have the required properties, and the ribonucleic acid (RNA) and di-ribonucleic acid (DNA) molecules are the prime examples [29] (in recent attempts to create artificial cells, a peptide nucleic acid or PNA was found to be an informational polymer analogous to DNA with the potential of controlling genetics, metabolism and containment of such artificial bodies [89]).

At the output or "action" end of information-driven interactions in the biomolecular domain are the proteins, manufactured by ribosomes, which are in themselves proteins. Proteins also are linear molecules, but since their purpose is chemical action and not information storage and transport, there are no restrictive conditions concerning the bonds between their own components. Indeed, this bonding contributes to the folding of these macromolecules into highly complex 3-dimensional entities with a clearly defined surface that governs their interaction with the medium in which they are immersed, and an interior that has a well-defined structure with specific properties and functions. Proteins are chains of hundreds or thousands of amino acids, but there are only 20 different kinds of such components; again, it is the *order* in which these building blocks are assembled and the length of the chain that determine the properties and functions of a particular protein. The number of possibilities for putting together a chain of N amino acids to form a protein is 20^N, again a ridiculously large number, but only a tiny fraction thereof (tens of thousands) are biologically active proteins.

In Sect. 1.4 we already referred to the DNA molecule and its informational structure. Let us now expand on this topic and briefly discuss the fundamental informational processes involved at the basis of all living organisms. A nucleic acid is a chain made of four kinds of molecular subunits, or nucleotides. Each nucleotide consists of an organic base that characterizes the nucleotide (adenine (A), thymine (T), guanine (G) and cytosine (C) in DNA; in the RNA thymine is replaced by uracil (U)), and sugar and phosphate units that are the same for all nucleotides of the chain. The link between neighboring nucleotides is a chemical bond between one sugar and one phosphate unit, thus marking a spatial direction on the molecule (Fig. 4.9a). Although we implied above that neighboring nucleotides do not interact with each other, hydrogen bonds are formed between cytosine and thymine (or uracil) and adenine and guanine, which in this way form complementary pairs. As a result, a free strand of nucleotides will collect with the help of certain enzymes individual complementary partners from the medium, string them together, and thus catalyze the formation of a "negative copy" of the original nucleic acid (Fig. 4.9b). If this negative is then "reversed" in a similar process, a copy of the original nucleic acid will have been produced. Actually, the DNA molecule is a double strand of complementary nucleotide chains (the famous "double helix"). If zipped open, DNA-synthesizing enzymes in the medium read the DNA template and bring in the right bases (in the form of triphosphates; two phosphates are released to provide the necessary bonding energy) – as a result, two new strands will emerge: The DNA has self-reproduced.

The synthesis of proteins is a complicated process, of which we will only present an oversimplified picture to point out some informational aspects. First, the transformation of hereditary information into proteins is a process that is common to all terrestrial life forms, practically the same in viruses, bacteria, plants, animals and humans. There are several fundamental correspondences involved in the relevant information-driven interactions. First, to each one of the twenty amino acids corresponds one *triplet* of successive bases of the DNA (in reality, it is an intermediate "messenger RNA" that carries a transcription of the DNA code). Such a triplet is called a *codon* (informationally, the equivalent to a "word"); since there are four letters (ATGC) in the DNA code (AUGC in RNA), there are $4^3 = 64$ codons. The universal relationship codon–amino acid is called the *genetic code*.[†] When the messenger RNA transports the information (a string of codons) to a ribosome, this molecular unit carries out the actual synthesis of the protein, successively pairing each incoming codon with the corresponding amino acid (collected from the medium), and stringing the latter together. Note that the speed of this process will be influenced by the availability, i.e., the actual concentration, of each relevant amino acid in the medium. There are "punctuation"

[†] There is some inconsistency in the nomenclature. The term "genetic code" is *also* used to designate the genome as a whole, i.e., the succession of bases in a DNA strand (see Sect. 1.4).

Chemical structure of the nucleotides

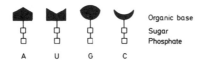

Part of a ribonucleic acid (RNA)

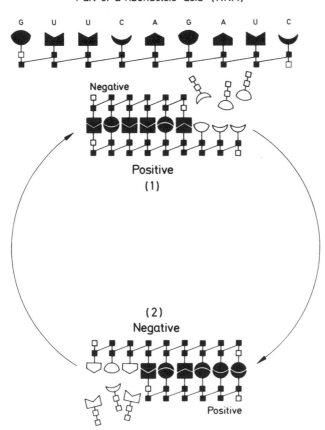

Fig. 4.9. (a) Scheme of the structure of a nucleic acid. Each geometric symbol corresponds to a base. The cross bonds of sugars to phosphates introduce a left–right asymmetry, which also defines a beginning and an end of the molecule. (b) Scheme of a nucleic self-replication process: (1) production of a negative copy, (2) reproduction of the original positive pattern. From *Küppers* [64], with the author's permission

codons, which tell the process where to start and where to end. A remarkable fact is that since there are only 20 amino acids, there are many more
codons available than needed! In fact, there are three amino acids each one
of which can be paired to (produced by) six different codons, five are paired
to four codons and nine amino acids are paired to two. The "stop sign" cor-
relates with three different codons. Biologists are feverishly trying to find
compelling reasons for the existence of this degeneracy or redundancy (see
review by [55]) – it hardly can be an accidental feature. There is indeed evidence from computer simulation studies that the specific relation between
the 64 codons and the 20 amino acids (and the stop sign) in the code offers
the optimal adaptive advantage [41]. In other words, it may have emerged
during the last four or so billion years to allow sufficient scope for evolution
of life on Earth. There may be more to the genetic code than just pairing
codons with amino acids; for instance, the existence of synonymous codons
(redundancy) could influence the rate of protein synthesis.

The existence of symmetries and palindrome patterns in the code could
be another indication of purpose beyond that of a mere mapping table [68].
The degeneracy-6 amino acids can be considered being coded by two subsets,
one with degeneracy 4, the other with degeneracy 2; with this assumption,
there are 8 amino acids paired to four codons and 12 to two. This presents
some interesting redundancies, especially if one is interested in the digital
encoding of the genome. Remembering our discussion in Sect. 1.4, this tells us
that the algorithmic information content of the genome must be less than the
amount of its Shannon information. Several investigators have addressed this
question; one scheme was proposed recently [49] that uses a binary nonpower
number representation with base numbers $\{8, 7, 4, 2, 1, 1\}^{\dagger}$ (remember the
somewhat similar Fibonacci number series!) that, when used to label amino
acids and punctuation signs, reproduces exactly the genetic code degeneracy
for the 64 codons (with the above assumption of degeneracy-6 coding) and in
addition reveals some surprising palindrome symmetry features of the genetic
code (the coding of each base is not straightforward, though: it is context-
dependent according to some specific rules).

$\dagger$ Power number representations are the ones used in the familiar decimal, binary and octal systems. In a binary nonpower representation, a set of bases
$b_n, \ldots, b_1, b_0$ is used that are *not* powers of 2 (and which progresses *slower*
than integer powers – each b_n must be $\leq 2^n$), and a number is expressed as
$x_n b_n + \cdots + x_1 b_1 + x_0 b_0$, where the x_i are 0 or 1. However, one given number
may have more than one equivalent representation; in other words, nonpower
representations have a built-in degeneracy or redundancy. For instance, for the
González base given above, there are 64 different possible number expressions
$\{x_n\}$, but many do represent the same number: the decimal numbers 8 to 15
have four different representations each (e.g., 13 = $\{011100\}$ = $\{101001\}$ =
$\{101010\}$ = $\{011011\}$); numbers 7 and 16 have three each; numbers 1 to 6 and
17 to 22 have two each; only 0 and 23 have one each.

In addition to DNA and RNA, there are many other molecules that play a fundamental informational role. First of all, the DNA is irrelevant without a "reading machinery" (with proof-reading capability), which happens to be different in every cell type. For instance the most common way for cells to modify transcriptional rates is to change the abundance of components that bind to specific DNA sequences, thus rendering certain genes more expressed (or less expressed). Under such circumstances, protein concentrations are also "inherited" (passed on to daughter cells), assuring the "epigenetic" control of gene expression. Indeed, one can argue that a huge amount of genetic information is encoded in the form of the concentration of these proteins; it is this, and not the DNA itself, what really makes a skin cell different from a liver cell. There are other fundamental information processing systems in a cell, for instance the signal processing that occurs after a receptor on the cell has been "engaged." This can take a variety of ways, but in general the modus operandi is that the receptor first brings together an enzyme complex, which operates on another cytoplasm protein. This in turn operates on yet another protein, initiating a cascade of signaling that eventually results in some specific action (turning on or off a gene, secreting something, etc.). The signal processing is in general catalytic (i.e., truly an information-driven interaction), although in some cases it can result in state changes of the actual signaling mechanism itself (similar to what happens in a neural learning network).

At a level intermediate in complexity between the cell and the brain we should mention the immune system: It has memory and has the ability to adapt to environmental challenges; it can be reset; it can discriminate between very fine differences (including "self" vs. "nonself"); it includes interacting and communicating elements; and it can multiply parts of itself. The B and T cells of the immune system, responsible for the adaptive immune response, actively rearrange (and even actively mutate) their DNA in order to generate the wide diversity of antibodies and T cell receptors needed to deal with infections.

There are other organic molecules that also play an informational role in biological systems, which do not actually process information but serve as carriers of signals in multicellular organisms. In animals, their mission is to stimulate or alter the function of a certain part of the body, an organ, the brain as a whole or a certain brain center. In the previous section we already referred to neurotransmitters (serotonin, dopamine, norepinephrine) which transmit signals from one neuron to another across the synaptic gap triggering an impulse in the postsynaptic cell. There are substances (endorphins, melatonin, acetylcholine, etc.), also pertaining to the class of neurotransmitters, produced mostly in the hypothalamus, basal forebrain and brain stem, which are released into the blood stream to act as "general modulators" of overall brain activity (globally modifying synaptic potency – a sort of "volume control" of the brain), to influence the activity of specific processing centers, or to stimulate the endocrine system's secretion of hormones which regulate organ functions. They play an important role in the control of emotions (Sect. 6.4). Plants do not have a nervous system but their functions are coordinated by specific hormones, some of them as light as ethylene (e.g., [2]), circulating through their vessels and controlling key functions such as germination and fruit ripening. Outside information about the state of the environment, besides the direct effects of radiation and air temperature, humidity and chemical composition, are conveyed to many plants by *pheromones,* organic molecules released into the air by other plants. Pheromones are also an important external communications device for some animals, including higher mammals and humans.

4.5 Information and Life

There is a major question to be answered concerning the information carried by the DNA molecule. Information on *what* is it, really, and where does it come from? It is clear that this information provides the machinery needed to construct and organize an organism that can "beat the elements," as we stated in Sect. 4.1, i.e., survive and reproduce. Survive and reproduce in the particular real-time conditions of a certain environment, the environmental *niche.* So in a sense, there exists a correspondence between the characteristics – patterns! – of the environmental niche and the patterns of the genome: Somewhere a "grand" information-driven interaction has taken place (Sect. 3.5). And the mechanism that establishes this correspondence is not one particular molecular machine, but a whole collective statistical-deterministic process: *Darwinian evolution* – a process that involves both statistical elements and a common, deterministic direction in time.

Let me give you a comparative "negative" example. Astrology works! – It works, because of the peculiar way our brain operates (Chap. 6): Programmed to remember mainly the relevant information from among the continuous onslaught of sensory input, it tends to forget, or not store at all, information on irrelevant or inconsequential happenings. So when an astrologer (or the daily horoscope in our newspaper) makes a series of (totally random!) predictions, we will remember those that by pure chance did come true, tell them to our friends, write them down in our diary, etc. – and forget all others. A Gaussian distribution has been cut off asymmetrically, and the average value of the data (in this case, the true/false value of each prediction) has been biased away from zero – information that was true (just by chance) survived and propagated, false information died off. If we were to continue with this biased mode of storing in memory, our brain would build a totally false image of events in the environment and our own place in it. This would represent a "negative" evolution – one that could lead our race to extinction!

Back to our subject: When the genome is ever so slightly altered, especially during self-replication, by perturbations from the immediate chemical environment, radioactivity, cosmic radiation, maybe even by some deliberately built-in random error-producing action, a slightly altered or *mutated* information is passed on to the descendants, leading to a slightly differently built organism. If this perturbation process is strictly random, a proportion of the descendants will be better suited for survival and the rest will only exhibit small or irrelevant changes, or be worse off and eventually disappear. The result is that *on the average,* the surviving offspring will always be better adapted to the environment, even if the latter is slowly changing. The mutation process has to be sufficiently slow so that the slightly altered genes can be passed on to offspring without any second-order perturbations undoing the primary change. Most mutations in the genome are acquired by mistakes in the replication machinery, although other factors like mentioned above certainly can be a cause, too. The mistakes come about by the infidelity of the responsible mechanisms, leading to mismatched pairing of two bases (Fig. 4.9b). An erroneous duplication of a base and its insertion into the *replicated* strand is indeed a critical mutation, because it allows the organism the flexibility to mutate the *copy* without changing the necessary function of the original. The system can "play around" with copies until it comes up with something "really good": For instance, in the human genome there are countless examples of genes that are "nonfunctional" (replicas of older genes that are no longer used and in fact do not encode a functional protein), but which can be mutated at high rates without any detriment to the organism – but with the benefit that by chance a novel protein could be produced with some function that might improve survival (the downside, of course, is the production of a novel protein that is "bad"!). In short, this has been a very effective way to evolve: Multiply something that already works, and tinker with it! That way, the system is not starting from scratch trying to evolve a particular activity in a given direction (which does not exist – from where would the information have come?), it simply adapts something that it does not currently need. In other words, in biological evolution, purpose *emerges,* it is not imposed from "outside."

The whole process of environmental adaptation is somewhat equivalent to the "musical" example of a neural network discussed in Sect. 4.1 that adapts itself to detecting the complex tones that are frequently present in our acoustic environment. The difference is that in the latter case the adaptation happens during the early lifetime of one and the same organism and works not by elimination but by reinforcement (of synaptic connections), whereas in the case of Darwinian evolution the adaptation is a slow collective process involving the survival and the elimination of thousands of successive members of a given species. With our definition of information in Sect. 3.5 as the agent that embodies the univocal correspondence established between a pattern at a source and a change (pattern) in some recipient, we can view the features of a particular environmental niche as "patterns in a source," and the characteristics of a species as the "elicited change in a collective set of recipients," and view the whole process of Darwinian evolution as a "grand multifaceted information-driven interaction" between environment and life.

There is another fundamental aspect of evolution. Randomness must govern mutations to confront the unpredictability of environmental change. As mentioned above, mutations come about by the infidelity of the duplication mechanism. It is almost certain that this infidelity is "programmed" – i.e., a given mutation rate is not only acceptable, but highly desirable. As an example, some viruses (such as HIV) have extremely high mutation rates due to the infidelity of the reverse transcriptase enzyme (converting viral RNA into DNA). This is precisely what makes HIV so successful, because it can "adapt" (through mutation) to the host's specific immune response very quickly (the cost is that a huge fraction (probably $> 99\%$) of the viruses produced by infected cells are in fact dead viruses).

Despite the random processes involved at the molecular level, evolution as a whole appears to progress in a deterministic direction, driven by those "programmed" errors toward an ever-increasing diversity of species and ever-increasing complexity of the evolving organisms [34]. The self-organization of the individual translates into collective self-organization of the genus. Organisms will adapt to very slow changes of the environment, and when these environmental changes are not uniform in space and time they will lead to a diversification of niches, hence a diversification of species. The pace of evolution on Earth has indeed been driven by geologic and atmospheric change, except for catastrophic events that introduced discontinuities in the Darwinian process, such as mass extinctions. There were other less sudden biospheric discontinuities which led to unusually rapid change. Examples are the appearance of blue–green algae as an important source of atmospheric oxygen which promoted the appearance of nucleated cells (eucaryots), and the appearance of sexual reproduction which sped up enormously the pace of diversification.

All this is ultimately the result of information-driven interactions with increasingly specific and diverse purposes – a massive collective self-organization. There is no need to invoke any metaphysical intervention: Numerical simulations using simple models of ecosystem evolution convincingly show this trend of increasing complexity, too (e.g., [83]), just as the computer models of adaptive neural networks reveal emergent (unplanned) collective properties which appear to us as stemming from some "intelligent design" (robustness to damage and input errors, ability to recall a full image with only a partial input, flipping between competing outputs, etc.; see Sect. 4.3). Note that the gradually increasing organization of the biosphere in the course of evolution seems to defy the second law of thermodynamics – but remember that we are always dealing with *open* systems (Sect. 3.1) receiving free energy (solar radiation in plants, food ultimately based on plants in animals, heat from volcanic vents on the ocean floor) and delivering entropy (waste) to the environment. Increasing organization means decreasing entropy (per member), and decreasing entropy means gradually increasing information (see Chap. 5) – this is precisely a general characteristic of evolution! Indeed, the information that the human genome contains about its own environmental niche is vastly larger than the environmental information mapped into a bacterium. Of course, we are talking here of information as defined in Sect. 4.3 (pragmatic information); the amount in bits of Shannon information (describing the sequence of bases) only differs by a factor of ~ 1000 (see Sect. 1.4)!

It would be a mistake to assume that everything produced by actions of DNA is reflected in the DNA code itself. Much of the outcome relies on the subsequent self-organization of the assembled structures, even the self-organization properties of some inorganic matter, a process that is not preplanned but *emerges* exclusively from physical and chemical laws, making wholes that are more than the sum of their parts (Sect. 3.1). In other words, evolution most successfully exploits physical properties that are "out there waiting to be used" (beware of this horribly anthropomorphic language!). For instance, DNA carries the information to make certain proteins, but it does not necessarily carry the information on everything these proteins must do – that has emerged through eon-long trial and error processes. Insect-trapping plants that execute rapid and precise, even intricate, movements whose details seem carefully planned, are only exploiting complex elastic properties and instabilities of the material of its petals or leaves, which just need to be triggered by a simple mechanical action by the insect.

So far we have focused mainly on single species. But evolution is a collective process, and when we talk about an environmental niche we must include in it the other organisms which are sharing it. There is cooperation and competition, symbiotic relationship and predation – in other words, a whole *ecosystem* evolves, adapted to that niche. It is the information on a particular niche *and* its ecosystem that is expressed in the genes and the ensuing physiological or behavioral characteristics of a given species; similar niches can lead to similar characteristics in totally different species (compare certain external characteristics of Australian kangaroos to European rabbits!); this is called convergent evolution. Changes of the environment are not just of geological and climatic nature. Life itself alters the environment significantly. There are many examples in the Earth's history: The atmospheric oxygen is the result of photosynthesis in green algae and plants; most atmospheric CO_2 originates in plant respiration and organic decomposition (now also in industrial activity); practically all surface calcium deposits are of organic origin and one third of chemical elements of the Earth's crust have been recycled biologically; what powers our cars today (and pollutes our atmosphere) is solar energy captured in photosynthesis hundreds of millions of years ago. In short: Evolution is an eminently nonlinear process that affects itself. As we already implied in regards of Table 3.1, Darwinian evolution is an integral part of the evolution of the Universe.

Finally, how and where did life actually start? A lot is known today about the genome, its functions in an organism, and the consequences thereof. But how did Nature get to the "first genome"? What was the first macromolecular ensemble that could be called "alive"? What was the first information-driven interaction? In short, when did information appear first as an active participant in the affairs of the Universe Table 3.1? There is a lot of fascinating research on in this subject, but much is still very speculative. Relics of primitive organisms are difficult to come by, and experimentation with the proverbial "primeval soup" [77] is still far from revealing the appearance of any macromolecules or assemblies of molecules that would meet the triad of conditions for life mentioned at the beginning of in Sect. 4.1. Exploration of Mars and the Jovian moon Europa, both potential bearers of life forms in the past and the present, respectively, has not yet provided any positive evidence of extraterrestrial life.

The point of departure for the emergence of information-carrying biomolecules is most likely the ubiquitous pool of small organic molecules such as methane, formaldehyde, alcohol and even amino acids, as mentioned in Sect. 3.1, and some "appropriate" environment and energy source, as they indeed did exist in the oceans of the early Earth. But the transition from that pool to a cauldron of large polymers and the first RNA specimens is still a great mystery. Catalytic reactions on the surface of certain minerals are suspected to have played a role. Once those components were available for action (protein synthesis) in a supporting environment, it is not difficult to imagine how Darwinism kicked in and life evolved as we know it (e.g., [29]). Most likely, even in a favorable environment of the early Earth, many "false starts" happened until the first molecular conglomerates that could qualify as living systems materialized in a stable form and propagated. One particular requirement of any theory on the chemical origin of life is to resolve the many feedback or "chicken and egg" situations in biomolecular processes. For instance, to have a functioning DNA one needs a reading machine, but to make a reading machine one needs DNA; to manufacture ATP (adenosine triphosphate, the fundamental energy-transport molecule operating in all cells), one needs ... ATP! And, of course, to make DNA one needs ... DNA! Theories for the transition from prebiotic molecules to living systems abound (see the excellent reviews by *Küppers* [64] and *Davies* [29]), and range from life being the product of pure chance [80] to theories with life somehow written into the laws of Nature, perhaps even quantum laws [74] – decoherence permitting (see Sect. 5.6)!

Some key unanswered questions are (see [64]): When does information become an active factor in the chemical evolution of life? How was the information-carrying nucleotide sequence of the protogene selected out of the innumerable chemically (energetically) equivalent possibilities? How were these extremely large and complex molecules pieced together from small prebiotic molecules? Could fluctuations and the existence of physically governed "attractors" have played a role in the primeval soup, channeling chemical compounds toward the assembly of RNA-like structures? Did quantum processes play a fundamental role (besides that of governing the chemical bonds)? Does quantum information play a role? Despite all these unanswered questions many scientists, me included, believe that Darwinian evolution is inevitable in view of the particular chemical properties of the carbon atom (which are universal). Given the appropriate environment and the necessary time, it is through giant carbon-based molecules that information-driven interactions can appear, and life with them.

Let me end this chapter with a quote from a recent book by *Braitenberg* [20]:

> Living organisms contain information, are the result of information, pass on information, are information made flesh on the conditions that their environmental niches impose for survival. To read this information is to recognize the very essence of a living organism ...

I would go one step further and repeat what I stated at the end of Sect. 3.5: *Life is information at work* – information appeared (in its fundamental pragmatic form) when and where life appeared in the Universe (Table 3.1). It plays no active role in the inanimate physical world.

5 The 'Passive' Role of Information in Physics

The title of this chapter sounds outrageous. After all, at the end of Sect. 3.1, did we not refer to *Chaisson's* statement [24] that "the process of cosmic evolution is continuously generating information"? Remember the relation between entropy and information mentioned in Chap. 1? And what about recent papers (e.g., [6]) that even found their way into the public media concerning black holes swallowing information? Or statements in the literature like "... in order for the Earth to follow in its orbit, it must constantly receive information about the ratio mass/distance ... of all other gravitating masses" (p. 26 in [54])? Throughout our text we, too, have made a number of statements that seem to link information directly with physical processes. For instance, in the discussion of force fields we stated: "... the purpose of a field is to communicate information about happenings at its source to any distant recipient; if this recipient understands the meaning of the message, it will react accordingly (wood shavings will not understand the meaning of a magnetic field ...)" (Sect. 3.3). We even called the field a "local ambassador" of its source. In that same section we said: "Whenever two opposite electric charges annihilate each other, electromagnetic energy is emitted as a messenger informing all points of space that 'the game is over,' collecting the local field energy and carrying it to infinity." And concerning quantum interactions (Sect. 3.4) we stated that "virtual field quanta ... are ... the messengers in elementary particle interaction."

What I have tried to achieve with such statements, however, is to emphasize how much subjective, anthropomorphic or even teleological parlance abounds in the description of purely physical processes. Indeed, we cautioned the reader about this every time, especially in Chap. 3. For instance, in connection with Chaisson's statement about information generation in the Universe we asked: "Are the physical, nonbiotic evolutionary processes and respective interactions really *using* this information? Did this continuously generated information have a purpose and meaning at the time it was generated, or does it have purpose and meaning *only now* for the brains which are *studying* the Universe?" We closed Sect. 3.4 with the statements (referring to some of the above anthropomorphic metaphors that seem to point to an active role of information in physical interactions): "... What kind of information is this? It certainly is *not* something that makes the Universe tick –

forces, fields and energy do that. ... [We are] dealing with information *for us*, for our models, for our description, prediction or retrodiction of a physical system. ... Overall, the use of the word information ... [is] metaphorical: it [is] used as a tool in our attempts to comprehend the physical universe in terms familiar to our biological brain." Finally, remember that when we discussed quantum processes we stated at the end of Sect. 2.3: "Our classical brain is desperately trying to make classical models of quantum systems in which information ... is running around inside the experimental setup at superluminal speed, as well as backwards in time, to intercept the photon wherever it may be, and tell it what to do!" At the end of that chapter we alluded to "... our unavoidable urge to represent mentally what happens *inside* the quantum domain with images and paradigms, such as tagging an indivisible particle and following its motion, borrowed from our sensory experience in a classical, macroscopic world." Well, let me generalize this latter statement and make it applicable to *all* domains of physics, and speak of "our unavoidable urge to represent mentally what happens in the Universe with images and paradigms borrowed from our *sensory* experience." It is important to understand how some consequences of this may affect the description and understanding of physics. Stated in other words: We must clearly distinguish between what is due to the human way of thinking and acting from what is due to Nature alone. In the present chapter I will elaborate on this.

5.1 The Human Brain: Turning Observations into Information and Information into Knowledge

It probably is not known how scientific thought has emerged from a *biological*, evolutionary point of view. I am neither an anthropologist nor a science historian, but this will not deter me from speculating. At the beginning of Chap. 3 we mentioned a fundamental function of the brain of higher animals, namely the capability of determining correlations between environmental events and storing this information in memory. And in Sect. 4.3 we described the process of associative recall as a most basic process of neural information processing, in which the repeated perception of two (or more) near-simultaneously occurring events can lead to synaptic changes in such a way that, in the future, the perception of only one of these events will trigger neural representations of the other(s). The more relevant to the organism a set of correlated events, the more efficient the learning process (see Sect. 6.4) – if the experience is scary or gratifying enough it may even work in a one-time occurrence. This process has two key spin-offs in terms of brain function: 1. establishing an *order of time* if the events are not simultaneous, and 2. establishing *cause-and-effect relationships* between events that appear relevant to the organism. In animals, the time interval within which causal correlations can be established (trace conditioning) is of the order of tens of seconds[†] and decreases rapidly if other stimuli are present [53]; in humans it extends over the long-term past and the long-term future. All this involves the generation and transformation of event-specific spatio-temporal patterns of neural activity in the cerebral cortex – in other words, *information* as defined in Sect. 3.5 – engaging the working memory and setting long-term memories (see Sect. 4.2, Sect. 4.3 and Chap. 6).

In the discussion of the process of associative recall (see Fig. 4.5), we have implied that it is triggered by perceptional events, i.e., input patterns from the environment (sight, sound, touch, smell) or from some internal somatic function (metabolic state, posture, etc.). In Chap. 6 we will discuss the unique capability of the human brain: to recall stored information as neural images or representations, manipulate them, and restore modified or amended versions thereof *without any concurrent external or somatic sensory input*. The acts of information recall, alteration and restorage without any external input represent the *human thinking process* or *reasoning* [91]. In other words, as we already mentioned in Sect. 4.4, the human brain can "go off-line" [12] and work on its own output without any correlation with events occurring outside the brain in real time. Concerning the determination of cause-and-effect relationships, the ability of recalling stored information without external input indeed leads to the ability of ordering events in time and discovering causal relationships long after they have happened, or correlations between current events that are separated in time by more than the short-term memory span of just a few tens of seconds. This ability endowed humans with the capacity of *long-term prediction based on observations*. For some subjectively relevant complex events the brain developed the capability of building a repertoire of possible outcomes and their respective likelihoods on which to base behavioral decisions (see Sect. 1.3 and Table 1.1!) – we may call this whole process "building a *knowledge* base." Rudimentary agriculture and animal breeding were probably the first "practical applications" of this ability in the societal realm; the belief in divine beings who govern uncontrollable aspects of environmental events, human life and death may have been the first "practical manifestations" in the spiritual realm. It is clear that the chain observations → information → knowledge is intimately tied to the characteristics and functions of the human sensory system and the neural networks that developed to process incoming information. Short-term predictions in all animal brains are a sort of interface operation between sensory input processing and motor (behavioral) response; long-term predictions in humans require a much more sophisticated interface processing, which resulted in a substantial evolutionary increase of the number of neurons and their interconnections in the prefrontal and frontal areas of the human cortex (compared to our primate ancestors).

[†] With this statement I will have earned the wrath of dog owners and chimp trainers. I will come back to this matter in the last chapter!

The presentation of an event (the input pattern) triggers a specific neural activity pattern in the brain, which at a higher level may lead to a set of multiple patterns representing alternatives, one of which is chosen as the behavioral response (the output pattern) based on some associated likelihoods (the probabilities). If there are no alternatives, the input–output correspondence is univocal, the behavioral response "automatic"; from the Shannon point of view (Sect. 1.2 and Sect. 3.6), for an outside observer watching, say, an animal's response to an environmental signal, the information gain would be zero (the response is anticipated). If alternatives exist (e.g., a dog faced with three bowls with different foods), the brain must go through a decision tree based on probabilities (e.g., Fig. 1.4b) acquired through experience ("I like the one with raw meat") or need of the moment ("the bowl nearest to me is easiest to get at"). Only the human brain can develop complex, sophisticated decision trees based on its ability to recall and order information *without* concurrent external input (e.g., "I won't eat because I am on a diet"). The spatio-temporal distributions of neural impulses (Sect. 4.2) involved in these operations do not establish exact correspondences: Not every detail of the environmental system under consideration is mapped into the neural networks. As a matter of fact, as already anticipated in Sect. 4.1, there are filters appropriately tuned to prevent any irrelevant or redundant information from clogging the processing networks of the brain (in Chap. 6 we shall sketch how this process is controlled by the affective or emotion-related centers). In other words, the brain works with *models* that are only approximations of the so-called "reality" outside – this represents an irreversible compression of acquired information (Sect. 1.4). These models are built in learning experiences and amended in later experiences, in which errors can be corrected and details can be expanded. In animals these amendments occur mostly by chance through repeated exposure to external happenings and/or genetic (instinctive) guidance; humans have the ability, always related to the fundamental capability mentioned above, to *seek out* improvements for the mental images they already possess. This led to a human drive toward a systematization of observation and representation, to the observation of events that were not immediately relevant, and to a more "statistical" approach to information acquisition.

Much, much later, long past the last stage of human brain evolution, came 1. the introduction of systematic *processes of measurement* and the development of *instruments* to extract information from domains not directly accessible to the sensory system, by converting the original information into patterns ("pointer positions," Sect. 3.6) that can be perceived; 2. the *sharing of information* through linguistic discourse and documentation of environmental information in more permanent and unbiased memory banks outside the brain; and 3. the mathematical idealization of regularities, symmetries and trends in cause-and-effect relationships in the form of *laws of nature*.

What I want to emphasize here is the (rather obvious) fact that the evolutionary expansion of the higher-level processing networks of the human brain that make all this possible did not take place in order to satisfy the intricate requirements for scientific thought, but, rather, was shaped by an explosive expansion of requirements for planning tasks to assure survival in a rapidly changing environment under increasing competition from peers (and the consequent need to outsmart, not just outpower, groups of competitors). The systematization of observation and prediction came only very late in the development of the human race, and had to content itself with the use of an "old" information-processing machinery whose evolution was driven by more mundane factors. Is it then a wonder that we have this "urge to represent mentally what happens in the Universe with images and paradigms borrowed from our sensory experience"?

As mentioned in Sect. 3.3, we can imagine, like Maxwell did, the concept of an electromagnetic field as an elastic, potential energy-bearing, continuous medium filling all space, although all we can measure are effects (forces) on electric charges (matter!) placed in it. We can imagine extra dimensions, although we must do so by considering 3D or 2D "projections"; we can even imagine curled extra dimensions and the "quantum foam" talked about in current high-energy physics and string theory. We can imagine elementary particles consisting of tiny vibrating strings instead of the mathematically bothering zero-dimensional point particles. In doing all this we use what we have inborn: the geometric concept of good old three-dimensional Euclidean space! But when "we use what we have" and imagine a photon running through our "quantum pinball machine" (the Mach-Zehnder interferometer of Fig. 2.2 and Fig. 2.3), or if we follow with our imagination separately, one-by-one, the two entangled particles shown in Fig. 2.6 and Fig. 2.7, we are violating a quantum physics law that does not transcend into the familiar macroscopic domain. Heisenberg's uncertainty principle (2.1) does not allow us even *in principle* to tag and follow particles as if they were projectiles in the classical macroscopic world! A similar restriction can be found in the relativistic domain: We can well imagine something moving with a speed that exceeds the speed of light (remember the motion of the intersection point of two straight lines making a tiny angle, mentioned in Sect. 2.8). But we are prohibited from imagining anything material or any classical field perturbation to move with $v > c$ because we would run into energy problems or problems with the order of time in cause-and-effect relationships, respectively. There are other limitations that must be carefully taken into consideration; for instance, although we can imagine models with a continuous medium or, in the other extreme, point particles moving on continuous paths, we must not forget that such mathematical abstractions would require an infinite amount of algorithmic information for their description. In general, because of our "old neural machinery," scientific intuition is condemned to be based on what this machinery has really evolved for, namely, inter alia, creating biologically useful neural images from sensory exposure to the classical macroscopic world. Early examples of the consequences of this are the long time it took civilization to overcome the Aristotelian image of an Earth-centered celestial system, and the difficulty of scientists until Newton's times to understand correctly the effects of friction on the motion of bodies. Even today the so-called creationists (a politically powerful fundamentalist movement in the United States) dismiss as "pure theory" anything that cannot be observed directly with our senses (unless it is written in the Scriptures).

They do have a point, though (but please do not tell them!): When we reach down into the quantum domain with our measurement instruments, the information to be extracted ultimately must be accessible to our sensory system, which means that it will have to be classical. But even in *classical* physics we cannot disengage ourselves completely from the peculiarities of human brain function. Precisely, it is when we ignore this fact that preconceived ideas about an active role of information in physical processes tend to pop up.

In what follows we shall first examine the process of *observation*. I want to augment the meaning of this term in a rather bold manner by designating as "observation" a series of steps that must be taken concurrently:

1. A decision of *what system* is going to be at the focus of our attention (component bodies, boundaries);
2. A decision of what details to neglect, which is equivalent to introducing an approximate *model* of the system;
3. A decision of *what observables* to measure (which must be compatible with the degrees of freedom left in the model) and what *units* and *frame of reference* are to be used for this;
4. A decision about the *initial conditions* or initial state of the system (which may mean either a decision about the time t_0 at which the pertinent measurements are to be made if we have no control over the history of the system, or a decision of how to prepare the system in order to achieve the desired initial conditions).

After analyzing some key aspects of this process of "observation," we will turn to the discussion of the process of measurement per se and the formulation of physical laws. We are not interested in epistemology as such – our discussion shall be focused almost exclusively on addressing two questions: 1. where and how does information enter the picture, and 2. where does the human brain operating mode affect the response of a physical system under study.

5.2 Models, Initial Conditions and Information

Our brain has a remarkable "topological" ability to which we will refer in the next chapter: the ability of "binding together" the often horrendously complicated signals which belong to one and the same object, received at different and often mutually distant regions of the retina. How this is done by the neural networks of the visual system is still to a large extent a mystery (though we have an idea *where* in the brain the corresponding operations are performed; see Fig. 6.2); a little more is known about how the brain binds together stimuli received on different places along the basilar membrane which belong to the same complex tone (Sect. 4.1). Binding features together is a fundamental operation that allows us to mentally associate different sources of information with *one* complex entity in space and follow the pertinent changes in time – we can speak of a flock of birds, the planets of the solar system or the molecules in a glass of water. This is the operation that allows us to *define* a system, in which, out of an infinite number of possibilities, our brain designates a limited ensemble of objects as "the system under study." By the way, this is also the mental operation that allows us to acquire the concept of "coordinable sets," the operation of counting, the notion of whole number, and thus lay the foundation of number theory and arithmetic.

There are two types of ensembles: a group of objects over which we have no direct control, like the birds, the planets or the molecules, and objects like the billiard balls on a frictionless table which *we* assemble in the laboratory. The components of any ensemble are in interaction with each other and with the rest of the Universe (Chap. 3). At some designated initial time t_0, they are in a certain *initial state,* which either can be determined by appropriate measurements (although for the molecules this is unfeasible from a practical point of view) or, in the case of objects under our control, can be deliberately set. We want to find the mathematical algorithms that allow us to predict the state of the system at a later time or to retrodict its past. This is, in short, the main objective of physics – but note carefully: It does *not* include the case of the birds! The underlying reason is that birds entertain *information-driven* interactions with the environment and with each other, and these fall into the realm of biology (Chap. 3 and Chap. 4).

To accomplish the purpose of prediction or retrodiction simplifications are necessary for practical reasons, and a *model* of the system and its components must be introduced. For instance, in first approximation planets and billiard balls can be considered dimensionless mass points; the molecules of water, instead, can be considered as "smeared out" in space until a continuous medium is obtained. The introduction of simplified models requires neglecting many degrees of freedom of the system, i.e., requires an a priori exclusion of extractable information. If more details of the evolution of a physical system are desired, the models must be refined and some neglected degrees of freedom must be readmitted into the picture. For instance, in our examples each planet or billiard ball can be given form and 3D mass distribution, friction can be introduced, and in the case of the water, density fluctuations can be allowed, etc. In summary, just as our brain only works with simplified images of the perceived world (Sect. 5.1 and Chap. 6), the discipline of physics also works only with simplified models of a system – initially crafted as *mental* images and later expressed in some idealized mathematical form.

Measurements, however, are made on the "real thing,"[†] but the observables to be measured must always be compatible with the degrees of freedom of the model chosen for the system under study. Note the trivial fact that once a model has been established for a given system, the initial conditions (now expressed as initial values of the variables retained in the model) can also be *imagined* instead of determined experimentally or set physically.

Let us now turn to a set of classical material points in mutual physical interaction (Fig. 5.1). Out of the "grand system," we select a subgroup of N mass points and designate it as *the system under consideration,* shown in the figure. At some designated initial time t_0 its components are respectively located at coordinates q_k^0 (which need not be Cartesian but must be compatible with the constraints[‡] and must be referred to an inertial frame of reference (Sect. 3.2)) with associated ("conjugate") momenta p_k^0. These are the initial conditions of the system, and the multidimensional vector $\boldsymbol{S}_0 = \{q_k^0, p_k^0\}$ represents the *initial state,* defining a point in *phase space.* Material points outside the designated system are considered "the rest of the Universe." All material points are in mutual interaction and we assume that Mach's relation (3.1) holds for each interacting pair. Forces like $\mathbf{f}_r$, $\mathbf{f}_s$ are external forces on the system; $\mathbf{f}_m$, $\mathbf{f}_n$ are internal forces (Fig. 5.1); we shall assume for the time being that these forces all derive from potentials, i.e., are conservative (Sect. 3.2). Not shown in the figure are possible constraints, which are assumed frictionless and at rest. The associated forces (called reactions of the constraints), responsible for restricting the movement of a mass point, arise in unspecified interactions (e.g., a perpendicular elastic reaction on a surface or curve). Perfectly elastic walls (e.g., the boundary B in Fig. 5.1 could be such) are considered constraints. Such a classical mechanical system is said to be *closed* if there are no external forces acting on any of its material points and there is no energy exchange involving radiation (holonomic constraints, whose forces are external but do not perform any work, are allowed – black body electromagnetic radiation within the enclosure is neglected in our semi-qualitative discussion). Strictly closed systems, however, do not exist – except for the Universe as a whole (Chap. 3).

We stated in Sect. 3.2 that Hamiltonian mechanics allows the determination of the coordinates and momenta of a system of mass points in interaction as a function of time. In other words, it establishes a *correspondence* between

---

[†] I am not going to get involved into a discussion of the "reality" of a physical system. Let us just tentatively say that it is what different individuals agree upon – i.e., the result of a "committee decision" (philosophers, please forgive me!).

[‡] The constraints (e.g., forcing a mass point to move along prescribed curves or surfaces) are represented by functional relations between the coordinates (holonomic constraints) or their differentials (nonholonomic); the total number of independent coordinates (degrees of freedom) will thus be less: $3N - N_c$, where N_c is the number of constraints.

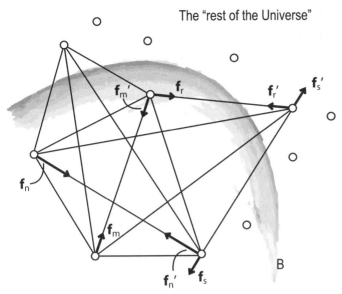

Fig. 5.1. Definition of a system of mass points and the interaction forces. $\mathbf{f}_m$, $\mathbf{f}_n$: internal forces, $\mathbf{f}_r$, $\mathbf{f}_s$: external forces. In a *closed* system, the boundary B shields the system from any external interaction. Strictly closed systems do not exist – the gravitational field is a geometric property of space-time and cannot be canceled in any given region of space

the *initial state* of the system $\boldsymbol{S}_0 = \{q_k^0, p_k^0\}$ at time t_0 and a *final state* $\boldsymbol{S}(t) = \{q_k, p_k\}$ at time t; a different initial state will lead to a different final state. This represents a mapping of one state into another, which can be represented with the use of a transformation operator T_t:

$$T_t \, \boldsymbol{S}_0 = \boldsymbol{S}(t) \,. \tag{5.1}$$

The integral transformation operator T_t embodies the algorithm of Hamilton's equations[†] applied to a given system of interacting material points. It represents the formidable *compression* of algorithmic information content that occurs through the pertinent physical laws, as mentioned in Sect. 1.4: The operator links the full description of the system at *any* time t, future or past, with that of the system at just *one* instant of time t_0. There are systems so sensitive to the initial conditions that for an infinitesimally small change of these, the transformed system $\boldsymbol{S}(t)$ may be radically different from the orig-

---

[†] The Hamilton equations are $\mathrm{d}q_k/\mathrm{d}t = \partial H/\partial p_k, \mathrm{d}p_k/\mathrm{d}t = -\partial H/\partial q_k$ where the Hamiltonian H is a certain function constructed from the system's generalized coordinates q_k, momenta p_k and the potential. For simple systems, it is equal to the total mechanical energy (kinetic plus potential): $H = T + V$.

inal one: $T_t\left(\boldsymbol{S}_0 + \delta\boldsymbol{S}_0\right) = \boldsymbol{S}\left(t\right) + \Delta\boldsymbol{S}$ where $\lim_{\delta S \to 0}\Delta\boldsymbol{S} \neq 0$. An example is the classical pinball machine of Sect. 1.2. Such systems are called deterministically *chaotic* (we cannot predict their evolution for merely "practical" reasons because we will never be able to determine physically the *mathematically exact* initial conditions required to repeat exactly the motion, but if we could, it would!). In what follows we will not consider chaotic systems, though.

Now taking into account the definitions given in Sect. 3.5 we realize that *when it is us who choose the initial state,* this physical correspondence would also represent an information-driven interaction: *We* set the initial state with the *purpose* to achieve a wanted final state, or with the purpose of finding out what the final state will be at time t. We have *used* the physical system for a mapping purpose; we can repeat the same process in the laboratory, in our mind or on a computer as often as we want, and for the same initial conditions (the same input pattern!) we will always obtain the same final state (the output pattern) – compare this with the discussion of Fig. 3.5!

I insist so much on the importance of the initial conditions of a classical physical system for several reasons (for a detailed discussion of all implications, see the article by *Bricmont* [21]). Above all, because this is one of the places where information finds its way into physics! Let us state it again: When *we* set the initial conditions of a physical system, whether it is by actually manipulating it or whether it is only in thought by manipulating a model of it, the system becomes *information-driven,* according to our definition in Chap. 3. We could even send a message by setting the initial conditions of a system of interacting mass points; the "common code" (Sect. 3.5 and Fig. 3.7) between sender and recipient would include the laws of interaction (operator T, relation (5.1)) and the instants of time t_0 (sending time) and t (receiving time). If on the other hand the initial conditions for any time t_0 are "given by nature," as is the case with the planets or the (inscrutable) case with the molecules in a glass of water, any purpose disappears, the correspondence between initial and final state is a purely physical one, and information plays no active role in the process. All that counts are the individual physical interactions between the mass points.

5.3 Reversibility, Determinism and Information

Given the initial conditions of a *natural* system at time t_0, Hamiltonian mechanics tells us how a system evolves after that initial time, but it does not tell us anything about the *reasons* of why it got to that initial state in the first place. Of course, we can use the Hamiltonian algorithm to follow the system backwards in time (see next paragraph) and retrodict its state until $t \to -\infty$, if we want to. But this will refer us only to "initial conditions" at some earlier times; it will never tell us *what process* got the system started in such conditions in the first place – it will never tell us anything about a physical cause. There will be exceptions in which we could infer something, though. For instance, if we follow back the positions of some small fragments orbiting the Earth, we may find that they all meet at a common point at some earlier time, suggesting for their origin the disintegration of a satellite (we will come back to this example again when we discuss thermodynamics). Tracing back in time the Moon's orbit (taking into account external perturbations from other planets, tidal energy dissipation, etc.) is indeed one of the techniques used by planetologists to determine its origin (these are examples how a physical system of interacting bodies is used *by humans* to extract information from the environment).

It is important to point out that the Hamiltonian evolution (5.1) of a closed classical system is *reversible* and that there is no preferred direction of time: The equations of mechanics are invariant with respect to the transformation $t' = -t$. If we invert the direction of *all* momenta p_k (or velocities) at the instant of time t and let the system run under the same interactions (conservative forces) forward in time (which in principle we *can* do – we cannot run a physical system backwards in time), each mass point will follow its original trajectory in reverse, and arrive after a (positive) lapse $(t - t_0)$ at the original initial position with reversed momenta. Another inversion of these momenta will then fully restore the original initial state. Basically, this is what one would see when one runs a film of the original evolution of the system "backwards," or, in our example above, what is computed when we retrodict the motion of the orbiting fragments of a satellite. We can express this graphically (Fig. 5.2). If we consider the operator T_t defined in relation (5.1) and introduce an operator R that reverses the sign of the momenta in the vector $S(R^2 = I$, the identity operator), the sequence of operations (inversion-evolution-inversion) would be represented by the product RT_tR, and we can write $RT_tRS(t) = S_0$. On the other hand, the operator T_{-t} running the system backwards in time is defined by the operation $T_{-t}S(t) = S_0$ (which means that $T_{-t}T_t = I$). We thus obtain:

$$RT_tR = T_{-t} \tag{5.2}$$

which demonstrates that the triple operation described above is indeed equivalent to a time reversal [21]. The important point (especially for later) is that there is *no privileged direction of time* in a closed system of interacting particles (valid only for conservative forces in the classical domain (Sect. 3.2), and in absence of weak interactions in the quantum domain (Sect. 3.3)). Note that all this also means that there are no real cause-and-effect relationships involved (something we already had anticipated in the previous paragraph). We cannot say "the cause of state $S(t)$ is the initial configuration S_0," because for a natural system there is no privileged direction of time (besides, the time t_0 is *our* arbitrary choice). But attention: If it is *us* who intentionally *set* the initial conditions, the character of the system changes and our intervention becomes the cause of every configuration that follows! Or if there

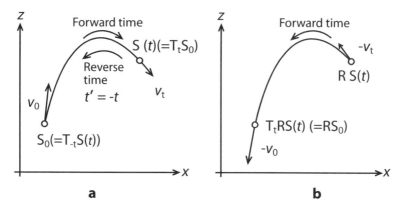

Fig. 5.2. System of one mass point in interaction with the Earth's gravitational field in vacuum. (**a**) Normal path from state S_0 to state $S(t)$. Going "backwards in time" means going from $S(t)$ to S_0; the point goes backwards not because we change the direction of the velocity, but because we count the time (abscissas) backwards. This cannot be accomplished physically. (**b**) What is feasible is to reverse the velocities (RS) and go *forward* in time to bring the system from S to its initial position – but it arrives with reversed velocity. Yet another velocity reversal restitutes the full initial state

are some externally controlled events like the disintegration of a satellite in a collision with a meteorite (an external interaction in the past that had nothing to do with the interactions within the closed system), that event becomes the external cause for the system's *particular* evolution. All subsequent states $S(t)$ are then a consequence of the causative event. Note, in preparation for a later discussion, that in either case, natural event or human intervention, these subsequent states $S(t)$ following the causative event would constitute a very *special* (and tiny) subset of all *possible* states of the system at time t.

A Hamiltonian system with conservative interactions is *deterministic* and *predictable*. The first quality has to do with the equations of motion which are deterministic, not stochastic, and which assure that for the same initial conditions there is only one possible state at any given time t (in the classical case – transformation (5.1)). In such a system, predictability or, rather, the lack thereof has to do only with *us*, with our ability to observe, prepare a system, make measurements, analyze data and compute. Errors in setting initial conditions lead to probability distributions of possible evolutions of the system (as any marksman would know) – but the intervening interactions and corresponding physical laws still are strictly deterministic. What *is* possible in such a case is to make predictions about probability distributions (see *Bricmont* [21], who presents the hypothetical example of a system that is clearly deterministic but intrinsically unpredictable). As Bricmont also shows, all this has nothing to do with chaos. A closed chaotic system is not predictable and reversible because *we* (or any intelligent system) will never be able in principle to *exactly* determine or set the initial conditions, even if it is purely classical. Of course, there are external, physical causes, that introduce irreversibility to all systems; indeed, strictly closed systems simply do not exist because it is impossible to shield them from external gravitational fields which, however weak (like, say, the gravitational field of Sirius at Earth [15]), may suffice to "trip" the micro-state of a system into going along a completely different path.

Let us now consider a system of interacting mass points that is not closed but forms part of a larger one, as shown in Fig. 5.1. Such a system is subjected to *external* forces and in general it will no longer be reversible: Reversing all momenta or velocities of the mass points of the limited system at a time t will not bring it back to the same initial state at time t_0 after applying the transformations given in relation (5.2) – unless *all* external interactions, too, would follow *exactly* the same pattern in reverse. However, this condition could only be fulfilled if we had total information on and control of the external part as well, such as the velocities and positions of all external mass points (and eventually any incident radiation). This shows that any external perturbation, however weak, would make a limited classical system of interacting bodies *irreversible* (see also discussion of Fig. 3.5 in Sect. 5.3). For strictly reversible systems we said that a privileged direction of time does not exist; this obviously is no longer true for irreversible systems. We shall elaborate on this when we discuss macroscopic systems in general (Sect. 5.5).

5.4 Microstates and Macrostates

Historically, macroscopic thermodynamics was developed to provide a framework for the quantitative understanding, with prediction capability, of heat transfer, efficiency of thermal machines, chemical reactions, etc., but it ran into problems with systems far from equilibrium. Whereas it could provide algorithms to find the direction into which a closed system would evolve toward equilibrium, it had difficulties in dealing with the processes responsible for the transition and predicting the speed of their development. In particular, it could not resolve the question of how an open system such as a living organism could be maintained in a semipermanent state of consistently low entropy, far from equilibrium. There were some other problems such as the experimentally observed existence of small fluctuations in a medium in thermodynamic equilibrium. Boltzmann's statistical thermodynamics began looking at the "other end" of the spatial scale, working with a model of mass points – the molecules making up a fluid. Much of statistical thermodynamics is based on gedankenexperiments, most of which would be impossible to reproduce in the laboratory, but whose logical consequences are verifiable (as happens with the Mach relations, Sect. 3.2). Notice how the observer now participates in a more active role: not only as a model-maker and condition-setter, but as an "imaginator" of conceivable situations. Concepts such as "uncertainty" and "knowledge" appear in mathematically quantifiable form – and information makes its official appearance on the stage of physics!

Figure 5.1 depicted a system consisting only of a limited number of interacting material points. We must now turn to a more realistic case – the $\sim 10^{23}$ molecules of which normal objects in our daily environment are made. If we want to link a macroscopic description with the *microscopic state* given by the position and velocity of each one of an object's molecules, we must begin by crafting an appropriate macroscopic model. For this purpose, we divide the object in question (assumed in thermal equilibrium) into small volumes δV_k, and work with the inertial masses δm_k of these elements of volume *as if* they were material points of a system. But one has to be careful with this subdivision. Consider an element of volume δV_k which includes point r_k, and form the ratio $\delta m_k/\delta V_k$ to define the local density of the body. We cannot let δV_k go strictly to zero, as the mathematicians would like us to do, because we would run into the domain of intermolecular distances. In this domain the total mass δm_k contained in the element would fluctuate madly, depending on whether or not molecules happen to be inside, and the ratio would vary erratically between zero and infinity as $\delta V_k \to 0$. Furthermore, there is another absolute lower limits, namely Heisenberg's uncertainty relation (2.1): to remain in the classical domain, we must assure that $(\delta V_k)^{1/3} \gg \hbar/p$, with p the (temperature-dependent) average momentum of the molecules.

However, there will be a certain range – the macroscopic domain – in which δm_k is nearly proportional to δV_k (Fig. 5.3) and the ratio $\delta m_k/\delta V_k$ is independent of δV_k; for larger values of the volume elements macroscopic inhomogeneities of the mass distribution could again destroy this proportionality. In the region of proportionality we can define a macroscopic quantity, the local macroscopic density $\rho_k = \delta m_k/\delta V_k$ at the point r_k in question. This is what sometimes is called a "physical differential quotient," as opposed to a "mathematical differential quotient" for which there are no lower limit restrictions. Treating physical differentials as mathematical ones is, indeed, tacitly equivalent to *making a model* in which we replace in our mind the discrete system in question with a different system: a *continuum*. We are de facto extrapolating to the microscopic domain the average ratio $\delta m_k/\delta V_k$ obtained in a certain macroscopic domain (Fig. 5.3) in which its variations and fluctuations are negligible. Only with the introduction of this *model of a continuum* can we begin working with nondenumerable values of our observables and use differential and integral calculus in a mathematically legitimate way (see discussion in Sect. 5.6).[†] All this seems "old hat" to anybody familiar with the physics of, say, elastic and electromagnetic properties of matter, but it is essential to understand the connection between a physical system "out there" and the model we make of it "in here" (meaning the brain).

[†] This model can be driven to an extreme for single mass points: Their continuum density is a delta function (see footnote † on p. 99).

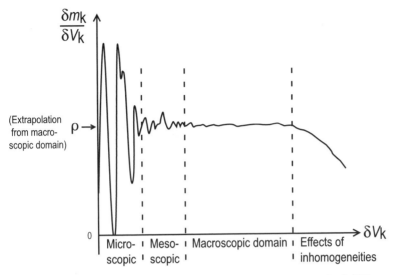

Fig. 5.3. Sketch of the value of the ratio (mass enclosed in δV_k)/δV_k, as a function of the dimension of δV_k, with indication of the different physical domains. Extrapolation of the curve to zero dimension defines the macroscopic variable "density" and allows the use of calculus in models in which the real medium is replaced by a (hypothetical) continuum

Let us consider the molecules of a gas contained in the left vessel shown in Fig. 5.4 (the tap between both vessels is initially closed). We assume that the collisions of the molecules with the walls (acting as constraints) and with each other are perfectly elastic (no total kinetic energy loss or gain in each collision), and that no other interactions, such as gravity, exist with the outside (an "ideal gas" as a "closed system"); we also neglect any radiation effects. This is a particularly drastic simplification because electromagnetic radiation is what "shapes" the velocity distribution of the molecules of a gas; however our assumption will not invalidate the basic points to be made below. Finally, we assume that the gas is in a macroscopic thermodynamic equilibrium state, which among other things requires that its density, defined at any fixed point in space (Fig. 5.3), is constant in time. Each molecule will be considered a *classical* mass point, which at any given initial time t_0 has a measurable position r_i and velocity v_i (measurable *in principle*, but not in practice!). If we knew the coordinates and velocities of *all* molecules, and had the necessary computer power, we could determine, using the equations of mechanics, the evolution of the system as a function of future time or retrodict its past, as we did earlier with a system of only "a few" points. But we cannot for practical reasons. What we do know, however, is that, regardless of how many molecules there are and what their distribution at any time is, this idealized system will follow a deterministic and perfectly reversible path

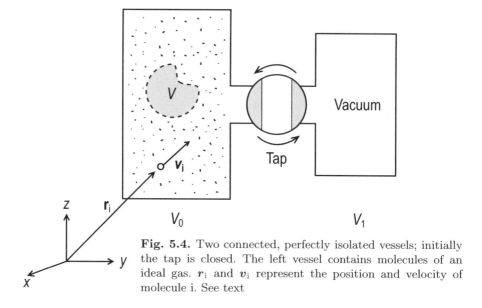

Fig. 5.4. Two connected, perfectly isolated vessels; initially the tap is closed. The left vessel contains molecules of an ideal gas. r_i and v_i represent the position and velocity of molecule i. See text

assuming states given by some operator T_t through relation (5.1). We have a random structure evolving deterministically – we cannot predict it because we do not know all initial conditions, but we do know the laws that govern its dynamics. In other words, this classical system does what the physical interactions between molecules and constraints "tell it to do" and it could not care less whether we are able to measure and compute it or not (careful with this anthropomorphic language!).

As mentioned before, our senses do not respond to individual molecules, but to the collective action of zillions of them. The same happens with macroscopic measurement instruments and the macroscopic devices we use to change a molecular system collectively (pistons, heat reservoirs, valves, etc.). This is why macroscopic variables are introduced such as density, pressure, temperature; integral state functions like internal energy and entropy; constraint-related parameters such as the volume to which a gas is confined; and the quantity of heat and mechanical work delivered to or received from the system representing macroscopic interactions with the outside. Like all observables, macroscopic observables are defined through a measurement process, and their values contribute to the overall description of the macroscopic state S_{macro} of a system. One objective of statistical thermodynamics is to link the macroscopic and microscopic descriptions of one and the same system, that is, link S_{macro} with S_{micro}, where the latter is represented by the vector $\{q_k, p_k\}$, which of course is crazily complex and time-varying even when the macroscopic state is constant. The macroscopic variables are mathematical averages of certain microscopic quantities (like kinetic energy, momentum flow, etc.), and are related to each other through the traditional thermodynamic laws which must appear as consequences or limits of the laws of mechanics that govern the components of the microscopic system.

Speaking of averages, we already mentioned in Sect. 1.1 that one can have many sets of data all of which have the same average and even similar standard deviations. They differ in their "microscopic" structure (their distribution) but are equivalent with respect to average "macroscopic" parameters. This is exactly what happens with the zillions and zillions of different microstates which all represent one and the same macro-system in equilibrium. The overwhelming amount of algorithmic information in a microsystem (the $6N$ real numbers $\{q_k, p_k\}$ of the N molecules at any given time) is condensed into less than a handful of macroscopic variables which are the ones that control all macroscopic interactions of the gas in question and to which our senses eventually can respond. However, in this compression huge amounts of information are irreversibly lost – everything on the details of the distribution of individual molecule parameters is gone; only those describing the collective behavior remain. Given any one of the microstates we can determine the macroscopic variables that define the "parent" macroscopic system. But given a macro-system in equilibrium defined by its macrovariables, we have utter uncertainty as to which one of the zillions and zillions of compatible microstates is at work. The remarkable thing is that it is possible to evaluate and express in mathematical terms this uncertainty or, more precisely, the changes in uncertainty when a macro-system changes from one equilibrium state to another. The macroscopic physical magnitude which measures this uncertainty is the entropy S – originally defined by Clausius in purely macroscopic thermodynamic terms. Given Shannon's measure of expected information H (or preobservation uncertainty; relation (1.3)) discussed in Sect. 1.3, it should be obvious now that there must be a relation between the two.

One particular task of statistical thermodynamics is to demonstrate the fact that, indeed, many different molecular distributions are compatible with the same macroscopic state of a system in equilibrium, and that as each possible microstructure evolves, the values of the macrovariables remain constant (see, however, the question of fluctuations discussed below). Interestingly, neither conventional nor statistical thermodynamics are able to say much about the actual time dependence of transient states of off-equilibrium systems, despite the fact that a microscopic system is governed by the deterministic Hamilton equations, which regulate its evolution in time. The main reason is that the time dependence of macroscopic variables is goverened by *mesoscale* phenomena (e.g., fluid flows, turbulence) which are collective and organized, but highly inhomogeneous dynamic processes proceeding deterministically but triggered and influenced by stochastic fluctuations (see also Sect. 3.1). This intrinsic difficulty with the treatment of time dependency presents substantial problems in biology, because living systems are, as mentioned on various previous opportunities, open systems in a long-term state of quasiequilibrium.

Let us return to our ideal gas in equilibrium enclosed in the volume V_0 of Fig. 5.4. At some arbitrary initial time t_0, the microscopic configuration of the $N (=\sim 10^{23})$ molecules is in some initial state $\boldsymbol{S}(t_0) = \{q_k^0, p_k^0\}$. Not all values of the q and the p are allowed: The walls of the vessel and total energy conservation within the closed system impose restrictions; this is why one calls these compatible microstates the "accessible" states. The fundamental postulate of statistical mechanics states that *all accessible microconfigurations of an isolated system in equilibrium are equiprobable.* Starting from some initial state, at a later time t the microsystem will be in a state $\boldsymbol{S}(t) = \{q_k(t), p_k(t)\}$, given by relation (5.1). We can view this dynamic evolution either as a single trajectory in the $6N$-dimensional phase space, or as a series of "snapshots" (a film!) of N points moving in normal 3D space. It is beyond the scope of this book to go into details, which can be studied in any book of statistical mechanics or thermodynamics, and we shall consider only the more intuitive "3D snapshot" approach. The reader should also be alerted to the fact that in all what follows, unless explicitly pointed out, energy considerations will be left aside; this means that our approach is oversimplified, but still correct insofar as the description of links to information (the main purpose of this book) is concerned. In the case of equilibrium, any possible initial spatial configuration (compatible with the spatial and energy constraints) is a *random* distribution, and so are the distributions that follow from it at any other time t. However, as long as we neglect any interactions with the outside, such evolving distributions are all *connected* in a deterministic and reversible way by Hamilton's equations to the initial distribution and among themselves. What is random is what is on the snapshot picture at any given time, but not its changes with time. Indeed, if we were to invert all velocities and launch the microsystem again, it would retrace all states taken earlier, as discussed in the previous section (for the time being, let us forget that we have no way of determining coordinates and related momenta or velocities for such large a number of particles).

Whereas the microstate of a thermodynamic system in equilibrium is changing continuously, the corresponding macrostate remains constant, by definition. Or does it? As a matter of fact, among the accessible configurations there are "bad" ones [21], still equiprobable and perfectly compatible with spatial and energy constraints, but which at some time or another would lead to a surprising fluctuation of macroscopic variables. Large fluctuations are extraordinarily rare – but they are not impossible! This means that a "perfect" equilibrium really never exists, simply because macroscopic systems are made of a very large, but still *finite,* number of particles! There will always be a "jitter" in the macroscopic variables, and this jitter may at one time or another reach noticeable proportions. At the mesoscale level, however, it is precisely this jitter that in an open system can be nourished by some external energy supply leading to macroscopic change or self-organization (e.g., the Bénard cells in a fluid that is being heated inhomogeneously, Sect. 3.1).

We can estimate the proportion of "bad" configurations. Let us consider a small but finite volume v contained within V_0 (Fig. 5.4). The probability p that any one *given* molecule of a random distribution would be found in that subspace v is $p = (v/V_0)$, regardless of where the others are. The probability that *all* N molecules are found in v is p^N; the probability that *none* of the N molecules would be there is $(1-p)^N$ – both horribly small numbers, but *not* zero! Notice that these are extreme cases of fluctuation, which, however, become less extreme, as the total number of components N gets down to "reasonable" values. For instance, if we throw ten billiard balls "at random" on a table, the probability that all ten will end up on only one half of the table is $(1/2)^{10} \cong 10^{-3}$, a rather "tolerable" value! But already for 100 balls, the probability decreases to 10^{-31}! In general, it can be shown that the probability P_{mN} for m of the N particles to be in v (the rest remaining in $V_0 - v$) is given by the binomial distribution:

$$P_{mN} = \frac{N!}{(N-m)!m!}p^m(1-p)^{N-m} \,.^\dagger \tag{5.3}$$

For given values of p and N, this function has a maximum P_{max} at $m = pN$; in our case $m = vN/V_0 = \overline{\rho}v$, where $\overline{\rho}$ is the average number density of the distribution. This proves that the most probable distribution is one with uniform density. But deviations are possible; what happens is that for the huge number of molecules in a gas under "normal" macroscopic conditions, as one moves away from the maximum, the function (5.3) decreases very, very rapidly toward zero; in other words, appreciable fluctuations are extremely rare. For $N \to \infty$, $P_{max} \to 1$. Herein lies the key to "macroscopicity": There is no real sudden transition from a microscopic system to a macroscopic one. A system with relatively few particles (say a table with billiard balls) will exhibit large fluctuations (in relative terms). The more particles are added, the smoother will be the behavior, the less frequent the larger fluctuations. In other words, there is no fundamental dynamic difference between the 10^{23} molecules of an ideal gas and the 10 billiard balls on an ideal table. The difference is for *us* observers: We can create (or perceive or imagine) average quantities in the first case that exhibit a continuous, smooth behavior (extremely small and infrequent fluctuations), but it would not make much sense defining (or trying to perceive, imagine) such quantities in the second case. Essentially the same happens with the curve in Fig. 5.3. For a "truly" continuous medium, which we can model as one with ever smaller (less massive) particles packed closer and closer together, relation (5.3) tells us that for $N \to \infty$, P_{mN} will be zero for all m except $m = pN = \overline{\rho}v$.

Note that one given number m does *not* represent just one accessible state: There are zillions of equiprobable states all of which have the same

---

† Remember that $N! = N \cdot (N-1) \cdot (N-2) \cdots 3 \cdot 2 \cdot 1$. Expression (5.3) represents, in general terms, the chance that an event of independent probability p will occur m times out of N tries ($m \leq N$).

number of particles m inside the volume v! But there will be "fewer zillions" when m is small and "many more zillions" when m is close to $\bar{n} = \bar{\rho}v$. What I am saying here, of course, is qualitative and utterly imprecise: We just cannot count states in such a mathematically unmanageable way. Helmholtz has created statistical thermodynamics by developing a "counting method" in which phase space is divided into small cells (physical differentials, small, but not *too* small) and regions in this space are identified that are compatible with the given constraints (walls, total energy conservation). The "number of states" is then expressed in terms of the total volume of cells available at any given time. Again, we will not enter into the details here.

How do we know if a given distribution is "bad" in the above sense of leading to a large fluctuation in macroscopic variables? If most of the N molecules are found initially inside the smaller volume v of Fig. 5.4, this would indeed be a bad (but possible!) distribution. But what if a presently "normal looking" distribution in V_0 was concentrated in v at some *earlier* time? (Think of the example of fragments orbiting in space that all came from one common origin!) This would mean that by reversing all velocities of this "normal looking" state, its molecules would assemble in the small volume v (like broken pieces jumping up to form a cup!). The same applies to an evolution forward-in-time. Let us pick one microstate *at random* out of all possible ones from an equilibrium distribution in V_0, and follow its particles. If within a "reasonable" time these particles all concentrate in the smaller volume v (or do some other shenanigan like all moving together in the same direction) the distribution indeed would be a "bad" one – but we had no way of predicting this until we actually followed the microsystem through time. The question is how long do we have to wait to see if a microstate picked at random reveals a "bad" behavior? Poincaré has estimated the average time (he did not use the words "bad behavior"!). He showed that any accessible random distribution will, after a certain average lapse of time (the Poincaré time – much, much longer than the age of the Universe (e.g., [21]) be identical to *any other* compatible distribution, picked at random)!† So now all equilibrium distributions are really the same thing – sooner or later each one has a chance of becoming "bad." Of course, in all of this we have assumed that there are no outside perturbations. But we have already stated that a completely isolated system does not exist – we cannot shield it from gravitational interactions, however weak these are. And a molecular system is chaotic in the sense that the slightest change in initial conditions may lead to a great difference in its evolution. This fact does not change most of our assertions – instead of evolving continuously, the microsystem will jump from one configuration to another in response to the minimal external perturbations, but such configurations are all equivalent, provided the energy transfer to or from the outside is negligible.

---

† The actual Poincaré time depends on *how* phase space is subdivided.

5.5 Entropy and Information

It is time to get things *moving* at the macroscopic level. Let us open the valve connecting the two vessels of Fig. 5.4. The gas in V_0 will stream into V_1 until both vessels are filled and equilibrium is reestablished. According to macroscopic thermodynamics, this process is an isothermal expansion (provided the complete system is fully isolated); the equilibrium pressure in the final state will be $P_f = P_0 V_0 / (V_0 + V_1)$, and the entropy will have increased by

$$\Delta S = S_{\text{final}} - S_{\text{initial}} = -kN \, \ln \frac{V_0}{(V_0 + V_1)} \, . \tag{5.4}$$

Notice that by opening the tap we have suddenly created what Bricmont calls a "bad" distribution of molecules, one of those huge fluctuations that would occur naturally only once in a Poincaré time! If we could stop the molecules in their track (after the expansion has taken place) and reverse their velocities, in absence of external perturbations they would all assemble again in the initial vessel V_0.[†] The uncertainty about their position has suddenly increased in relative terms: Before opening the tap we knew that the molecules were all in the volume V_0 (although we did not know where exactly each one was therein); after the expansion each molecule could be in *either* V_0 or V_1! (There is no change in the uncertainty about their velocities, because the internal energy of the system remained constant in this isothermal expansion.) We can quantify this increase in uncertainty in a simple way. Suppose the space in both vessels is divided into small elements of equal volume δV. The number W_i of accessible spatial states available to *one* molecule before the expansion will be proportional to V_0 (if δx is the error in position measurement, and we set $\delta V = \delta x^3, W_i = V_0 / \delta V$); after the expansion, we have $W_f \sim (V_0 + V_1)$. In other words, W_i and W_f are the number of equiprobable (equilibrium) alternatives available to one molecule before and after expansion, respectively. According to relation (1.4), the Shannon expression of average uncertainty H *per molecule* will be $H_{\text{initial}} = \log_2 W_i$ and $H_{\text{final}} = \log_2 W_f$, respectively. For N molecules, assuming that there are no physical restrictions as to how many molecules can occupy the same element of volume, the individual H values are additive and we obtain a change in total uncertainty (change in Shannon entropy):

$$\Delta H = N \log_2 W_f - N \log_2 W_i = -N \log_2 \frac{V_0}{(V_0 + V_1)} \, . \tag{5.5}$$

[†] It is easy to verify with this example a statement made before: The dynamics and duration of the transient stage during which the gas streams through the tap into the initial vacuum will depend on many "circumstantial" factors affecting mesoscale processes, like the diameter of the tube, the (otherwise) constant temperature of the gas, turbulence, etc. Thermodynamics per se does not make statements about time dependence in off-equilibrium situations – only about *what* state of equilibrium will eventually be reached.

Note that the change ΔH is independent of the type of molecules and the actual size of the volume elements δV – what is important is the proportionality of the number of possible states W and the total volume. The actual increase of the number of accessible states as a result of the expansion is enormous; if we pick any state at random, the probability that we would catch an outgrowth of the original state (the one before the expansion) is extraordinarily small, given by relation (5.3) for $p = V_0/(V_0 + V_1)$ and $m = pN$. Comparing (5.5) with (5.4) we realize that in our example the thermodynamic entropy is related to Shannon's entropy by

$$S = k \ln 2\, H, \tag{5.6}$$

and that an increase of uncertainty by 1b is equivalent to an entropy increase of $+k \ln 2$, as indeed anticipated in Sect. 1.3. Equivalently, the gain of 1b of information about the microstate of a system represents a loss of $-k \ln 2$ in entropy. Going back to the number of accessible states in our example, we can write

$$S = kN \ln W, \quad \text{or per molecule: } s = k \ln W, \tag{5.7}$$

the latter of which is Boltzmann's famous expression of the entropy in statistical thermodynamics (embossed in his tombstone). Missing in relation (5.7) is an additive constant, which in classical thermodynamics is taken to be zero (the entropy of a gas at $0°$). The key point is that (5.7) is *valid in general;* W is the total number of accessible states or configurations taking into account not only the spatial states as we have done in our example, but also all possible energy states of the molecular components of a gas compatible with the total internal energy of the closed system. And here is where the difficulties of statistical thermodynamics began in the early part of the 20th century: Boltzmann's method of counting (to determine W) had to be revised to take into account quantum-mechanical properties of atoms and molecules. This required including the internal energy states of the particles, spin, and a subdivision of phase space in which the minimum volume element is $\delta V \delta p \sim$ h (Planck constant, Sect. 2.1). For instance, depending on their spin, different statistics apply because of certain restrictions of how many particles can share the same state in phase space (Fermi–Dirac statistics for half-integer spin particles like protons and electrons which are subjected to Pauli's exclusion principle; Bose–Einstein statistics for integer spin particle like photons and deuterons). Again, it is not our purpose to go into details, which are documented in many textbooks (e.g., [38]).

Historically, Clausius' entropy, defined exclusively in terms of macroscopic variables and macroscopic thermodynamic interactions, represents the unavailability of "usable" energy (work) in a system; Boltzmann's statistical entropy measures the uncertainty (lack of information) about the microscopic state of an ensemble of classical particles. Both entropies are numerically identical for a system in equilibrium. In addition, different *coarse grained* entropies can be defined for one and the same system, depending on how phase space is subdivided. All these statistical entropies have to do, though not explicitly, with a relationship between the system and the observer; they involve a brain's cognitive state (Sect. 3.6 and Chap. 6), although the intervening quantities are all defined by the system itself. They all coincide on the average for a system in equilibrium.

What I have described above are expressions related to the second law of thermodynamics taken from a statistical, microsystems point of view: When a fully isolated physical system is initially off-equilibrium, it will change until reaching a final macroscopic equilibrium state in which the number of microscopic states W which are compatible with the macroscopic conditions (spatial confinement and internal energy) is maximum, which, since they are equiprobable, also means that *maximum uncertainty* about the microscopic state of the system has been reached. In other words, a closed system evolves toward a maximum number of equally probable accessible microstates. This number W (which vastly exceeds the number N of particles in question) determines the degree of uncertainty (or the average information to be gained in a full determination of the microstate) defined by relation (1.4), and thus also the entropy according to (5.7). In summary, the state of equilibrium of a closed system is indeed a state of maximum entropy. Among the accessible states, there is a tiny fraction (see relation (5.3) and related discussion) that give deviations (fluctuations) of the macroscopic variables (this latter fact cannot be extracted from purely macroscopic thermodynamics – see remark on fluctuations in a "truly" continuous medium in Sect. 5.4). To decrease the entropy of a system, i.e., the number of possible microstates, the system must be "opened" to interactions with the outside (exchange of work, heat, matter).

Let us go back to Fig. 5.4 and do a thoughtexperiment with the equilibrium state achieved once the gas (of classical molecules) has expanded into both volumes V_0 and V_1. We close the tap again, preventing molecules from V_1 going into V_0 and vice versa. Nothing happens macroscopically as long as we keep the total system isolated from external interactions. Now we open the tap again – still nothing happens macroscopically, because the gas is in the same macroscopic state on both sides. Yet, regarding the microscopic state, *we know* that the $N_0 = NV_0/(V_0 + V_1)$ molecules from V_0 will "expand" into vessel V_1, and the $N_1 = NV_1/(V_0 + V_1)$ molecules from V_1 will "expand" into V_0. Consequently, according to relation (5.1), there should be an increase in conventional entropy given by

$$\Delta S = \Delta S_0 + \Delta S_1 = -kN_0 \ln \frac{V_0}{(V_0 + V_1)} - kN_1 \ln \frac{V_1}{(V_0 + V_1)} > 0. \quad (5.8a)$$

This micro ($\Delta S \neq 0$) vs. macro ($\Delta S = 0$) contradiction is called the *Gibbs paradox*. When the molecules initially in V_0 are *distinguishable* from those in V_1 (e.g., two different species, two isotopes of the same element, or molecules of two chemically identical compounds that have opposite helicity or symmetry), the above relation is indeed *correct*, and there is an entropy increase when both classes are mixed (notice that ΔS is independent of the gases in question). This is physically understandable: There are more states available for each distinct gas after the expansion than before; it is extraordinarily unlikely that, after reaching equilibrium the mixture would "unmix" and each component would assemble again in the vessel from whence it came.

Now, what happens, when *indistinguishable* molecules mix? An increase in total entropy like (5.8a) would be absurd! This paradox is usually solved by saying that properties of matter change discontinuously: Two molecules are either identical or they are not – either we have a process of mixing or not. If we do not have a process of mixing, we must apply a "counting rule" for identical molecules which is different from that for two different species (e.g., [38]).[†] Let me propose a different argumentation, which uses the concept of information [95]: What we have done in the case of mixing identical molecules is a thoughtexperiment in which we have opened the tap and followed *in our mind* molecules from one side to the other *as if they had been marked*, i.e., identified with some immaterial pattern ("this molecule is from V_0," "that molecule is from V_1"). This is similar to what we were tempted to do mentally in our discussion in Sect. 2.3 with quantum particles (but were prevented by Heisenberg's uncertainty principle)! In the present classical case, however, we are *not* prohibited from mentally marking the molecules, but if we do, we must take into account the fact that we are *decreasing* the uncertainty of the system (we now know from which side each molecule came). This should be equivalent to a decrease in entropy. We can calculate the amount of entropy extracted or, what is the same in this case, the information added to the system.

Let us rewrite relation (5.8a) in the following way:

$$\Delta S = kN \ln 2 H_{\mathrm{m}} \tag{5.8b}$$

in which $H_{\mathrm{m}} = -p_0 \log_2 p_0 - p_1 \log_2 p_1$. The $p_i = V_i/(V_0 + V_1)$ ($i = 0, 1$) are the probabilities that one molecule picked out at random of the final ensemble originated in vessel V_i. H_{m} is the Shannon entropy (1.2) per molecule, a measure of the a priori uncertainty in regards to its origin (being from V_0 or V_1), or, equivalently, the average information (in bits) to be expected from a measurement leading to the identification of the marker of that molecule (see Sect. 1.3) (notice the "coarse graining" involved: we are not interested in any microstate, or in the initial state of the molecule in phase space; what matters is just whether the molecule came from vessel V_0 or from V_1). If the marker is a *physical* feature that could be identified through a physical (force-driven) interaction, the molecules in V_0 and V_1 are truly different species, and H_{m} and ΔS represent true physical entities – the entropy increase in such a case

---

[†] In the calculation of the number of accessible microstates, we must divide each W_i by the corresponding value of $N_i!$, in which case it can be shown that, indeed, $\Delta S = 0$.

is legitimate. Now, if originally the molecules are *indistinguishable* and we mark them *in our mind,* we are adding prior information to the system, i.e., we are decreasing the average uncertainty per molecule by an amount ΔH which happens to be exactly $\Delta H = -[-p_0 \log_2 p_0 - p_1 \log_2 p_1] = -H_{\mathrm{m}}$.[†] So in our mental experiment the steps are as follows: 1. We mark the molecules as they diffuse from one vessel to the other; this adds average information to the system by the amount $N\Delta H$ and, according to (5.6), decreases the total entropy by $\Delta S = -kN \ln 2H_{\mathrm{m}}$. 2. We let the system mix; when equilibrium is reached, according to (5.8b) the entropy will have increased by $\Delta S = kN \ln 2H_{\mathrm{m}}$. The net change of total entropy is zero!

There is a subtle point here. It may not be difficult to accept that a process like "painting patterns on molecules" in our mental experiment will change the Shannon informational entropy, because it has to do with introducing alternative microconfigurations (in particular, see end of Sect. 3.6). But is it not intriguing that this change also implies a change in the "good old thermodynamic" entropy, that entropy which appears in Carnot cycles, efficiency of heat engines, etc.? To address this question we should focus our attention on the *interactions* involved in the process. In our gedankenexperiment, in which we tagged molecules in our mind and followed them as they diffuse into each others domains, we have *information-driven interactions* (Sect. 3.5) involved in the purposeful setting of markers (patterns). And the setting of the patterns is an external action, i.e., we have *opened* the system and extracted entropy from it (deposited and stored information in it). Having opened the system, we should not be surprised that a change in pattern alone (a change in Shannon entropy) indeed will call for a change in physical entropy (e.g., Boltzmann or statistical entropy). This even applies to small populations of macroscopic objects: If I take a bowl full of colored marbles and distribute them into separate bowls according to their color, I will obtain a system of lower entropy than if I had merely distributed the unsorted mixture into those same bowls. I can do the ordering only if I can entertain an information-based interaction with the marbles (besides the physical interactions with my fingers): In a dark room it would not work! A nice and tidy room is a system of lower physical entropy than the same room with the same objects scattered around at random. If an ordering process happens naturally in a system without the intervention of a living system (i.e., exclusively on the basis of force-based interactions leading to self-organization, Sect. 3.4), the system must have been open to a high-grade/low-grade energy exchange and the entropy decrease must have occurred at the expense of a larger entropy increase elsewhere (see our discussion in Sect. 3.1). If, on the other hand, the ordering process is the result of information-based interactions (the ordering of colored marbles, the tidying of a room, the synthesis of a protein), the entropy decrease must occur at the expense of a greater entropy increase

---

[†] This is easier to see if both vessels are of equal volume ($p_0 = p_1 = 0.5$): In that case the average information added per molecule is exactly 1b!

within the interaction mechanism itself (living system or a human-designed machine) that mediates the information-driven interaction (pattern recognition mechanism, metabolism of the brain, radiated heat, etc.).

Why are Shannon entropy and thermodynamic entropy mathematically related (identical)? Because the more we know about the microstructure of a system (the lower our uncertainty), the more we could anticipate (predict) actual fluctuations and somehow exploit them (e.g., decreasing the volume of a gas without doing work on it by deforming the boundary surface to fill void spaces). In classical thermodynamics entropy is a quantitative expression of our "frustration" of not being able to use in the form of mechanical work all the internal energy that is contained in a system. In statistical thermodynamics, entropy is a quantitative expression of our "frustration of not knowing." In the case of the molecules of a macroscopic body we do know about the existence of an "almost infinite" number of alternatives (this number is a relative entity: it depends on how we divide phase space into small cells). All these alternatives are compatible with what we obtain in macroscopic measurements, but we are unable to determine which is *the one* actually realized. If we could catch all the information and, at least for one instant of time, *know* that particular microstate of the system, all uncertainty would have instantly been removed, and the entropy of the system would have collapsed to zero, just as the Shannon entropy or average expected information collapses the moment we look at the result of a pinball machine operation, a coin toss or dice throw (Sect. 1.3, p. 22). As we mentioned in the preceding section, in the classical case we could follow the evolution of the microstate of the system using the algorithmic information that Hamilton's equations provide (e.g., relation (5.1)), and the entropy would remain zero as long as the system remains strictly closed. However, there is a hitch in this train of thought: As in the gedankenexperiment with marking the molecules in an isothermal expansion, the measurement operations to determine the position and velocity of the molecules require opening up the system, i.e., allowing interactions with the outside! There will be accompanying processes: The observer (a "Maxwell superdemon") must be equipped with appropriate energy tools for these actions (for the "conventional" Maxwell demon,[†] see [67]).

Finally, let us assume that somehow we can kick the superdemon out, once it has provided us the information, and take it up from there. The initial statistical entropy would be zero in all respects, even the physical one, because it is as if we started from a zero absolute temperature with all molecules frozen in place (zero energy, zero entropy), and then gave each molecule the right velocity (total energy = internal energy, entropy still zero). The en-

---

[†] Maxwell demon: a tiny intelligent being that operates the valve in Fig. 5.4. Initially, both volumes of gas are connected and in equilibrium; the demon lets pass "fast" molecules in only one direction and "slow" molecules in the other, with the end result that heat will be transferred, apparently without requiring any external work, from one vessel to the other.

tropy should remain zero if deterministic mechanical laws govern unhindered the evolution in time of that initial microstate (relation (5.1)). But let us remember that there are no totally closed systems. The microstructure would quickly jump from the evolved known state into any other nearby accessible state as the result of unavoidable interactions with the rest of the Universe – the tiniest perturbation would suffice. The result will indeed be an extraordinarily rapid loss of our information, with the entropy very quickly attaining the nominal equilibrium value. Note an important aspect: If the system is already in equilibrium, the number of accessible states is maximum and any tiniest perturbation will throw the microsystem into a "neighboring" distribution (Sect. 5.4), but the probability of hitting a "bad" distribution and maintaining it a sufficient time to notice any major macroscopic effects is extraordinarily small (relation (5.3)). If, however, a system is not yet in equilibrium, like our expanding gas in the preceding section, the probability of jumping into a configuration that is still bad from the point of view of the final system (gas in both volumes $V_0 + V_1$) is much, much greater – indeed, during the expansion process, additional limitations to accessible states still remain in effect, such as the finite speed of the molecules. The unavoidable increase of entropy and uncertainty or loss of information about the microstructure, and the concomitant increase of the number of accessible states in a closed system out of equilibrium, all point to a clear direction of time, called *the thermodynamic arrow of time* [66]: Reverse evolutions are not prohibited, but they would be extraordinarily rare and short-lived. Irreversibility (Sect. 5.3) is thus a statistical concept.

5.6 Physical Laws, Classical and Quantum Measurements, and Information

One of the succinct ways to formulate the Second Law of thermodynamics is to say that a closed system always evolves toward a state of maximum entropy. We have a peculiarly anthromorphic way of thinking about (and teaching) physical laws: We imagine a system "to do what the physical laws tell it to do" or "to obey" the physical laws, just as a good citizen obeys the laws imposed by some higher authority. In other words, we tend to view time evolution as a physical system choosing, from among infinite different imaginable alternatives, only that path which is compatible with the physical laws. Yet it is a trivial remark to say that inanimate systems have no way of obtaining and processing information on possible alternatives and that they cannot make any decisions! It is the fundamental physical interactions which have certain properties that reflect regularities and limitations to the evolution of a system in independence of its initial conditions and constraints (e.g., Sect. 3.3, Sect. 3.4 and Sect. 5.2), and it is *us human beings* who have the ability to identify and document these regularities and limitations through our own information-driven direct or instrument-aided interactions with Nature.

For instance, when we say that a Hamiltonian system of mass points "takes a path of minimum action," as the Lagrange "principle" states,[†] we are obviously not implying that the system looks around, probes the terrain (in configuration space) and chooses that path for which the action integral turns out minimum (which is exactly the mathematical procedure to derive Lagrange's "principle" from Hamilton's equations). Rather, it means that we, humans, have discovered a certain integral (whose integrand is a function of the system's instantaneous, observable, state) which, if calculated along the actual path of a system, yields an extreme value (minimum) compared to that obtained for any other path (compatible with initial conditions and constraints) that we can *imagine* for the system. Likewise, when we say that the molecules of a gas move in a way so as to maximize the entropy of a closed system, it really is the other way around: We are able to find a function of certain thermodynamic variables called entropy, whose value tends to a maximum and remains there when the molecules move the way they do. In its statistical/informational interpretation, entropy expresses the degree of incompleteness of an inquiring observer's knowledge – but the molecules do not care at all about information.

Physical laws are algorithms that express natural limitations of variability in mathematical terms, reducing enormously the algorithmic information needed to describe the evolution of a system (see pertinent discussion in Sect. 1.4, Sect. 3.2, Sect. 5.2), and enabling us to make predictions and retrodictions (Sect. 5.1). Notice that our brain can envision so many more alternatives – the marvel of a physical law is not what it "does" or "prescribes" but what our brains can imagine *could* happen but does not! Finally, let us always keep in mind that physical laws are *idealized* expressions, because they first require the formulation of a simplified *model* of the system under study (Sect. 5.2). To ask if mathematics and physical laws "exist independently" may be a valid question for philosophers to address, but it is an absurd one within the realm of pure physics. Nature behaves the way it does because of deterministic interactions (correlated changes) between its components; we are the ones mapping one set of configurations – patterns – into another and discovering ways to order the information obtained (the correspondences) so as to be able to predict or retrodict the behavior of the simplified models we have made of selected parts of Nature. And we are the ones who will either be satisfied about the match between prediction and further measurements, or dissatisfied and look for improvements of the models and pertinent physical laws.

[†] $\int L dt$ = minimum, where $L = T - V$ is the Lagrangian (T is the total kinetic energy, V the scalar potential of the forces involved). More correctly, we should express this principle in variational form as $\delta \int L dt = 0$, where δ represents the variation of the integral when taken along a path infinitesimally proximate to the actual course between the initial and final points.

Physical laws are based on measurements. What is measured are the observables, each observable being defined through the process of measurement per se (really, a class of measurements). There are many books written on the subject (e.g., [105]) and we will only address a few fundamental points in the classical and quantum domains. We have referred to the measurement process already on several previous occasions (e.g., Sect. 1.2, Sect. 2.1, Sect. 3.6); it represents an information-based interaction *par excellence:* the establishment of a correspondence between a pattern "at the source" (the system to be measured) and a change at the "receiving end" (e.g., the position of a pointer, the track in a recorder, the state of the brain of a person counting objects) which would not occur in absence of this interaction. The univocal correspondence between the pattern being measured and the change evoked embodies the (pragmatic) information obtained in the measurement (see Sect. 3.5 and Fig. 3.8). The ultimate purpose of a measurement process is always to evoke certain neural patterns in a brain that we call "new knowledge." For the relationship between pragmatic information and Shannon information I should refer to Sect. 3.6.

One important point is the fact that the amount of information obtained in any measurement is always expressed in a *finite* number of bits. Of course, this is guaranteed by the ultimate limitations imposed by accuracy (in the classical domain) or by Heisenberg's uncertainty relation (in the quantum domain). Let us illustrate this statement with an example. Suppose we want to measure the position of a point along the coordinate axis x (distance between the point and the origin O) with a binary ruler (Fig. 5.5). The (arbitrary) unit of length is segment U, and the a priori error or resolution is ε. The *physical* part of the process of measurement consists of the determination of the minimum-size segment into which the point falls; the binary label of the left marker is then taken as the result or "value of the coordinate" (top graph). The resulting binary number (1.010 in our example) also gives the algorithmic information in bits (ignoring the binary point); it is easy to verify that it represents the number of steps (number of times the unit has to be apposed to reach the point, plus the number questions "is the point in the right half or left half" of binary subdivisions, that must be answered). An important fact is that, for a constant error ε, the number of bits is independent of the unit (e.g., [54]). To show this, in the bottom graph of the figure we have chosen a unit that is one half the original size; note that in the final figure (10.10) the binary point changes, but not the order of the digits (this happens because the relation between the new and the old unit is $\frac{1}{2}$). In general, if one changes the unit, the number expressing the value X of an observable will change according to the well-known rule $X' = Xu$, where X' and X are the numerical values of one and the same observable expressed in the new and the old units, respectively, and u is the value of the old unit U measured with the new unit U'. What perhaps is not so trivial is the fact that the information *content* of any measurement is always finite

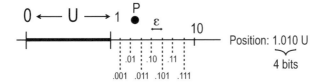

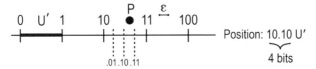

Fig. 5.5. *Top:* the process of position measurement in binary form. *Bottom:* case of a unit of length half that used in the top graph. Given an a priori resolution ε, the algorithmic information content of a measurement result is independent of the unit chosen and always finite (quantum information can have infinite information content, but it is not accessible to measurement – see text)

and always the same, regardless of the unit chosen (provided the resolution of the measurement is the same).

In our original definition (Sect. 3.5) of information-driven interactions, we relate a pattern at a source to a change at the recipient. But any change is relative – we required that it be with respect to some standard initial state to which the mechanism must be reset every time after an interaction has taken place. In the case of a measurement, this requirement means that the instrument (the pointer settings) must be *reset* to the same initial state. What does this mean in the case, say, of the measurement of length (Fig. 5.5)? In essence, the standard initial state is achieved with the measurement of the unit of length, that is, the *calibration* of the instrument. There is another important issue. We said early on that in information-driven interactions, energy is required but it is not a representative factor (Sect. 3.5). The same applies to a measurement; the amount of required energy varies from case to case [112]. However, a fundamental question arises: Is there something like a *minimum* amount of energy? (We are in the classical domain, so quantization is not an issue.) We have the expression (5.6) of thermodynamic entropy per bit, namely $s = k \ln 2$; so is there a minimum amount of energy associated with the creation or erasure of 1b of information? This question has been much discussed in connection with Maxwell's demon [67], and it also has basic implications for computer technology. The usual assumption made is that this minimum expenditure or gain of energy can be derived from the classical definition of entropy, $\Delta S = \Delta q / T$, where ΔS is the increase of entropy of a body (e.g., an instrument) which has received a small amount Δq of heat in a reversible process at temperature T. Thus one considers the quantity $\Delta E = T \Delta S = kT \ln 2$ as the minimum thermodynamic energy required to remove an uncertainty of 1b (i.e., to gain 1b of information) from a system in thermal equilibrium at temperature T.

Let us turn now to the quantum domain. The information to be extracted in a measurement ultimately must be accessible to our sensory system, which means that at the observer's end it has to be classical (Fig. 3.8 and Sect. 5.1). However, the other extremity of the measurement device is sticking into the quantum world, where things behave very differently, and what is translated into classical signals of finite informational content (the "position of pointers") is nonlocal quantum information of potentially infinite information content. Between the two extremities of a measurement apparatus there is a chain of processes (physical interactions) that transform some specific quantum process (e.g., scattering of a particle) at one end into a macroscopic change (e.g., pointer position) at the other. Understanding this concatenation and identifying where exactly the quantum–classical transition occurs is a most difficult subject, still very much under discussion. We can only make a few basic comments and refer an interested reader, who should be well versed in quantum mechanics, to the recent literature (e.g., [47]).

First, we must return to the wave function, mentioned only in passing on p. 40 in the introductory Sect. 2.1 (for details, see for instance [11]). It is a complex function $\psi = \psi(\mathbf{r}_k, t)$ of coordinates and time and represents the state of a system of particles in configuration space (defined by all $3N$ coordinates – the degrees of freedom[†]). Basically, it has the same properties as the state vectors discussed in Sect. 2.1 and Sect. 2.4, except that sum operations must be replaced by integrals. Two of the fundamental properties of the wave function for *one* particle are:

$$\int_V \psi^* \psi \, dV = p_V \qquad \text{and} \qquad \int_\infty \psi^* A \psi \, dV = \langle A \rangle \qquad (5.9)$$

where p_V is the probability of finding the particle in volume V and $\langle A \rangle$ is the mean or expectation value of the observable represented by operator A (obviously, the wave function must be normalized: $\int_\infty \psi^* \psi \, dV = 1$). Any operator representing an observable that has a classical equivalent is a function or differential operator; the eigenfunctions are still defined as $A\varphi_k = a_k \varphi_k$, but this now may be a differential equation. The eigenfunctions φ_k must form a complete orthogonal set; any state ψ can then be expressed as a vector in "regular" Hilbert space: $\psi = \sum c_k \varphi_k$, with the c_k complex numbers as in (2.5), but where the sum over k now extends to ∞. The coordinate operator for each axis s is just plain x_s and the conjugate momentum operator is $p_s = -i\hbar \partial / \partial x_s$; the quantum operator equivalent to the Hamiltonian in one dimension is therefore (Sect. 5.2) $H = p^2/2m + V = -(\hbar^2/2m)d^2/dx^2 + V(x)$, whose eigenvalues are the

---

[†] There are internal degrees of freedom like spin and polarization which must be added explicitly.

energy levels E_k of a one-particle system: $H\psi_k = E_k\psi_k$, i.e., the solution of:

$$-(\hbar^2/2m)d^2\psi/dx^2 + V(x)\psi = E\psi. \tag{5.10a}$$

This is the time-independent one-dimensional *Schrödinger equation*. Any linear combination of solutions of (5.10a) is also a solution, i.e., a legitimate superposed state of the system. Taking into account the second equation in (5.9), the average position of a particle is $\langle x \rangle = \int_\infty \psi^* x \psi dV$ and the average x-component of the momentum $\langle p_x \rangle = -i\hbar \int_\infty \psi^* \partial\psi/\partial x dV$. Using the solution of (5.10a) for a free particle ($V = $ const.) and inserting pertinent quantum expressions in the definition of standard deviation (see Sect. 1.1), one can prove that Heisenberg's uncertainty relation (2.1a) is indeed satisfied.

It is clear that the solutions of (5.10a) depend on the form of the potential $V(x)$ and on the boundary conditions. In most cases, the values of E_k form a discrete set, maybe infinite but denumerable (the *essence* of quantization). Examples for simple configurations (electrostatic and elastic potentials, square well, periodic potentials in a solid, etc.) are discussed in most books of QM. An important fact shown by these examples is that there are configurations, like the low-quantum number energy levels in a deep potential well, that are very insensitive to the details of the potential far from the equilibrium points, i.e., they pertain to systems that are robust with respect to influences from "the rest of the Universe." Examples are the stable bounded structures like atoms and certain molecules – the "attractors" mentioned in Sect. 3.1.

The evolution in time of the wave function is given by an evolution operator T_t (similar to (5.1)): $\psi(t) = T_t\psi(t_0)$. The operator must be unitary (Sect. 2.5) and its limit for $t \to t_0$ must be I. This equation can be converted into a differential equation in the limit $t \to t_0$: $\partial\psi/\partial t = \partial T/\partial t\psi$. The key now lies in the time dependence of the evolution operator T. This requires a *new* QM principle, which reads: The time derivative of T is *proportional* to the Hamiltonian operator (e.g., see [11]). It can be shown that the constant of proportionality must be $-i/\hbar$, otherwise (5.10a) would not be the limiting case of the dynamic equation for static conditions. This principle leads us to the *time-dependent* Schrödinger equation:

$$i\hbar\, \partial\psi/\partial t = H\psi. \tag{5.10b}$$

This equation is also valid for a system of N material points. The important thing is that, as emphasized in Sect. 2.8, the state of any system, therefore its wave function, must include every particle that we designated as part of the system – but it must also include from the very beginning any external particle that may at one time or another interact with the designated system's components. In other words, equation (5.10b) is valid only for truly *closed systems*.

With the above new QM principle for time-dependent systems, the principle of superposition is broadened: Any linear superposition of time-dependent physical states is also a physical state. The more particles are included in a system, the more superposed states, solutions of (5.10a), are possible, i.e., the more complexly entangled a system can become. But also, the more opportunities will appear for external influences to break up the entanglements and "mess up" mutual phases. When this happens, it will start behaving classically: Superpositions of the type (2.19) (mutually entangled) will become of the type (2.18) (individually superposed states but mutually disconnected), and the latter in turn will break down (collapse) into a *statistical mixture* of zillions of independent states. This is in essence what is called *decoherence* [117]. If this were not to happen with the ensembles of particles in macroscopic bodies, they would be able to retain entangled properties – individual objects would loose topological identity and form, macroscopic variables could not be defined, our brains would be in a permanent state of multiple personality, and the "macroscopic" world would be very strange (see beginning of Chap. 3, and the paragraph below). The obligation to include all interactions with the outside in the wave function presents many conceptual difficulties, because, contrary to the classical situation, in the quantum domain we cannot isolate a system *even in thought!* In the classical case, we do not have such limitation in principle, although we must recognize that in practice we cannot isolate a system from gravitational interactions (which as discussed in Sect. 5.5 leads to the perturbation of any deterministic evolution of a classical microsystem and the increase of the entropy even if at some instant of time we possessed total knowledge of its microstructure).

Let us consider the sketch of Fig. 5.6. We have a Qbit which is exposed to a measurement apparatus to determine the status in which it is found once "activated" (Sect. 2.7). This means that at one extremity of the measuring instrument there must be an interaction site with the Qbit; at the other extremity there must be a macroscopic feature whose change can eventually be "perceived" by an observer. Inside the apparatus, there are causal physical links that transform a quantum process into a macroscopic effect. And, finally, there is the "rest of the Universe." Let us examine this arrangement quantitatively step by step. At the quantum end of the apparatus, certain quantum properties like the original superposed state of two components $|\psi_{sys}\rangle$ (relation (2.15)) may be preserved after the interaction, leading to a superposed, composite, system $|\psi_{app+sys}\rangle$. As we move up the chain further toward the "pointer," more and more particles and interactions will be involved, and some quantum properties will be lost due to the action of external influences – however weak. The complete wave function $|\psi_{rest+app+sys}\rangle$ is no longer a pure state that can be represented as a linear superposition in which the terms are neatly associated with either one or the other eigenstate of the original Qbit. If we ignore the unavoidable interactions with "the rest of the Universe," we

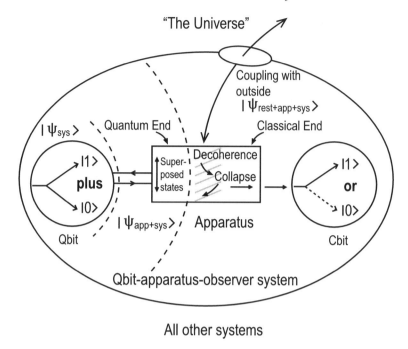

Fig. 5.6. Sketch of the interacting components in the measurement process of a Qbit in superposed state. At the "classical" end of the measuring apparatus, there is a pointer or, in the case of the Schrödinger cat paradox, a cat in a box set to die if one of the two alternatives occurs. See text

would indeed come up with paradoxes like the famous *Schrödinger cat*,[†] which is found in a superposed state identical to that of the Qbit (same coefficients) of being alive and dead at the same time – until the observer looks into the cage and the cat's wave function collapses into one state or the other [51]. Well, the collapse happens somewhere before, not as a single collapse but as a breakup of the state into a mixture of "zillions of pieces."

One of the fundamental postulates of QM tells us that the original Qbit, if measured a second time, would yield exactly the same experimental result, i.e., be found in the same eigenfunction in which it was left after the first

---

[†] In short, this paradox is as follows (for details, see, e.g., [51]): A Qbit is prepared in an eigenstate and measured. The output of the measuring apparatus activates a mechanism that kills the cat if, say, the state of the Qbit is $|0\rangle$, and does nothing if it is $|1\rangle$. Now the Qbit is prepared so as to assume a *superposed* state, and measured. An observer opens the box to see if the cat is dead or alive. If the original superposed state is preserved throughout all stages of the measurement process, the cat would also end up in such a superposed state: dead and alive at the same time. Since this is never observed, if found dead, when did the cat *actually* die?

measurement. We may be tempted to view this as a retropropagation of the collapse to the original system; however, taking into account the discussion in Sect. 2.8, especially in relation to Fig. 2.5, there is only one "grand" wave function involved from the very beginning of the process: $|\psi_{\text{rest+app+sys}}\rangle$. Let us repeat the statement made in a footnote way back on p. 40 of Sect. 2.1: The quantum wave function is not to be confused with an ordinary traveling wave, whose amplitude describes intensity or energy density, which carries energy and momentum from one point to another at a finite speed – none of these properties apply to a quantum wave function. The wave function *is* – it is an abstract nonlocal representation of the state of a system that does not propagate from one place to another, although from it one can obtain information on the time-dependent probability distribution of moving particles (represented by propagating wave packets). Let us face it again: A time-ordered causality applies only to macroscopically measurable effects.

We mentioned information – let us cast this in more specific terms. From the first of equations (5.9) we realize that $\rho = \psi^*\psi$ is the *probability density*[†] for finding the particle at the point in question. According to equation (1.1), $I = -\log_2 \rho$ is a measure of the "novelty value" of actually finding the particle there, and from (1.3) $H = -\int \rho \log_2 \rho \, dv$ is the Shannon entropy of the system. Just as the wave function from which it is derived, quantum information is nonlocal and propagates with infinite speed; but we just cannot think about it as "running around" from one place to another in the system under consideration (Sect. 2.3)! The wave function is an abstract mathematical bridge in *our* description of a quantum system, linking the initial conditions that *we* set for a system *we* have defined with the responses that *we* obtain with *our* measurement instruments – and these instruments as well as ourselves participate as an *integral part* of the system. Notice the multiple emphases! This is the formulation of what is called the orthodox Copenhagen interpretation of quantum mechanics. There are other interpretations and formulations which, however, we will not discuss here.

Let us now apply a train of thought similar to the one used in the last paragraph of the previous chapter concerning entropy. We *could* imagine that the dual status of a binary superposition of the original Qbit is preserved throughout the measuring apparatus, including the (classical) part of the instrument, leaving the pointer molecules (e.g., Fig. 3.8) or, what is equivalent, the cat molecules collectively in a dual superposed state: $|\psi_{\text{rest+app+sys}}\rangle = c_0|\varphi_{\text{alive}}\rangle + c_1|\varphi_{\text{dead}}\rangle$, where the φ are the "Universe's" wave functions with a living and a dead cat, respectively, and c_0 and c_1 are complex coefficients of the original Qbit state (except for a different phase factor). What would we see, if we looked at the pointer or into the cat box? Nothing

---

[†] A very important QM operator is the *density matrix*. If a state $|\psi\rangle$ is defined in terms of its components in a given orthonormal set of eigenfunctions $|\psi\rangle = \sum c_k \varphi_k$, the density matrix elements are $\rho_{ik} = c_i^* c_k$. Note that the probability density $\rho = \sum \rho_{kk}$ = trace of density matrix.

surprising: They would be in one *or* the other position. Why? Because the duration of a coherent dual superposition in a system of so many components would be extraordinarily short, just as happened with the duration of an artificial zero entropy ("all known") microstate of the molecules in a gas! The pointer would point, or the cat would die, perhaps just barely a Planck time later than it would have done in a normal experiment at room temperature! Why do macroscopic bodies have a stable, macroscopically well-defined form and surface (if solid), why do "chunks" of compact matter appear (Sect. 3.1)? Quantum superpositions and entanglement are in principle allowed to extend over large populations, bringing a "fuzzy" picture into the scene. But, like our gas, which in principle is not prohibited from suddenly assembling all in one corner of the vessel, the probability or duration of such a weird or "bad" state (Sect. 5.4) would be incredibly small! In the measurement of a quantum system, the external influences unavoidably "rattle" the measurement apparatus where it is most vulnerable – at the transition from a few to many components. All this was very qualitative; a rigorous discussion of decoherence for advanced readers can be found in [115].

An interesting situation arises if the macroscopic change in the instrument is *autoreversible,* e.g., is erased automatically (out of human control) after some time. If it is erased in such a way that in the interval between the actual measurement (interaction with the original quantum system) and the erasure there was *no way,* not even in principle, to extract information on that macroscopic change, no decoherence would occur in the original system, even if the process left the system in a different quantum state (i.e., a transformation occurred (Sect. 2.5)). What happens in that case is that "to extract information" in the interim physically would require to *expand* the measuring system by including a second fully classical apparatus that interacts with the "classical end" of the original one (Fig. 5.6), and to include it in the total wave function. In that case, the measurement process must not be considered as finished until the *second* instrument has shown an irreversible change (of course, this could be extended into a whole chain of successive measurement systems). If there is none, regardless of what happens in between, there was no measurement (no information-driven interaction, no evoked macroscopic change), only a transformation of the original quantum system (like the "splitting" at a half-silvered mirror). Table-top experiments have been made (e.g., [113]) that show how the erasure of a nonobservable change restitutes the original characteristics of a quantum system.

Let me address once more the question of "understanding" the behavior of a quantum system. "Quantum intuition" is represented more appropriately by the capacity of *accepting facts* than by the dexterity of creating mental pictures – which unavoidably will have to be drawn from the classical realm. *Landau* and *Lifshitz* [65] state it clearly:

> quantum mechanics occupies a very unusual place among physical
> theories: it contains classical mechanics as a limiting case, yet at the

same time it requires this limiting case (our brain images) for its own formulation

(the statement in parenthesis is mine – see also what I say at the end of Sect. 2.9). For sure, the molecules of the "pointer" of the instrument and the photons that activate our retina also belong to the quantum domain, but it is their collective action, represented by macroscopic variables, that leads to stimulus-specific neural patterns in our brain, i.e., represents the extracted classical information.

An appropriate way to end this chapter is with two quotes and a personal summarizing remark. One quote is from *Davies* [28]:

> the answer to the question 'Why is the universe knowable?' is un-
> known. The dazzling power of mathematics to describe the world at
> a basic level continues to baffle us. The ability of a subset of the
> universe (the brain) to construct an internal representation of the
> whole, including an understanding of that basic level, remains an
> enigma. Yet a comparison of computational and natural complexity
> surely provides a clear signpost for the elucidation of these age-old
> mysteries.

The other quote is perhaps quite related. *Tegmark* states in one of his papers [109]:

> An accurate description of the state of the universe appears to require
> a mind-bogglingly large and perhaps even infinite amount of infor-
> mation, even if we restrict our attention to a small subsystem such
> as a rabbit. ... most of this information is merely apparent, as seen
> from our subjective viewpoints, and ... the algorithmic information
> content of the universe as a whole is close to zero.

Based on what I have discussed in this chapter and the two preceding ones, I would like to conclude with this remark: When it comes to the physical, nonbiological world, the information is only in our heads – it does not actually *do* anything out there!

6 Information and the Brain

This book is not a treatise on the brain. Yet the reader will have realized that references to brain function, particularly human brain function, have appeared with insistence in all preceding chapters. As a matter of fact, they have played a prominent role in many of the points I was trying to make. As stated in the Introduction, "Information is Us," and since this is a book on information it is only logical that the "Us" – our brain – be a key protagonist in it. But only in some specific, limited, albeit fundamental respects – anyone interested in the details of brain anatomy, physiology and general functions should consult the vast medical and clinical literature available (e.g., [23, 59]). As a bridge to the preceding chapters, the ultimate purpose of this last chapter is to illuminate three interrelated issues, intimately related to the concept of information (see Table 1.1!): 1. a brain's transition from the state of "not knowing" to the state of "knowing"; 2. a brain's transition from "not knowing" to "imagining"; and 3. the drives and motivations that lead a brain into one transition or the other. I am convinced that a better understanding of the neurophysiological and neuropsychological mechanisms responsible for these processes will lead to a more objective understanding of the concept of information per se and of such diverse but related concepts as the "novelty value" and Shannon's entropy (Chap. 1); the limits imposed by nature to understand quantum systems (Chap. 2); the design and use of models in physics, and entropy as a quantitative measure of uncertainty concerning the microstate of a system (Chap. 5); and the role of a planner and observer in the measurement process, especially in the case of quantum systems (Chap. 2 and Chap. 5).

6.1 What the Brain Does and What It Does Not

As species evolved, information about the environment was gradually incorporated and stored in the genetic memory structures of the organism (Section 4.4). Very slow changes in the environment, of time-scales orders of magnitude longer than that of one generation of the species, could be incorporated in the genome by Darwinian evolution. But as the species became more complex and as the reaction to more and more unpredictable characteristics of the terrestrial environment became determinant of survival, the capability of ontogenetic adaptation during the lifetime of an organism became a fundamental requirement. When locomotion appeared in multicellular organisms some half-billion years ago, the number of relevant environmental variables to be monitored increased drastically, with the timescale of change down to a fraction of a second. It became necessary to absorb an enormous amount of information through increasingly sophisticated sensory system. Most of this influx of information is irrelevant but it carries embedded in an a priori unpredictable way those signals or patterns that are decisive for the organism's survival. The nervous system evolved to endow higher organisms with the capacity to detect, sort out, and identify relevant information contained in the complex sensory input, and anticipate and react appropriately to fast changes in the environment. To "react appropriately" also requires receiving and processing information on the organism itself, its overall metabolism and posture, position in and relation to the environment. In the course of this development, what started out as a simple environmental signal conversion, transmission and muscle reaction apparatus in cnidarians like jellyfish and sea anemones, evolved into the central nervous system (CNS) of higher vertebrates, with sophisticated input-analysis and response-planning capabilities. And, within the CNS, the brain emerged as the "central processor" carrying out the fundamental operations of monitoring and control of somatic functions, environmental representation, prediction of environmental events and the execution of a life-preserving and species-preserving behavioral response.

It is interesting to point out that the brain evolved in a quite different way from that of other organs. Instead of just expanding the capacity of existing neural networks, new structures with fundamentally different missions and processing modes were gradually added. This had a special evolutionary advantage: The basic, genetically programmed control of life-preserving functions and reactions could be maintained while the overgrowing structures could carry out new, more sophisticated information-processing operations and eventually subordinate the more primitive functions of the older structures turning them to the advantage of the new tasks.

There are some fundamental facts concerning brain operation, many of which already were addressed in Chap. 4 and in other chapters, which we shall summarize and take as the departing point for further discussion:

6.1.1 The brain works on the basis of information encoded in the form of the *distribution in space and time of electrical neural impulses* (Sect. 4.2). In general, the whole brain with its 10^{10}-plus neurons and 10^{14}-plus synapses (in the case of the human brain) is involved in this, although, as hinted before and as we shall examine in detail later, there are specialized modules or processing centers for specific tasks which, however, still may involve as many as 10^5 to 10^8 neurons each. We also said that each brain state is represented by one specific distribution, and that in contrast with, say, the microstates of a gas, there are no "equivalent" distributions that represent one and the same overall macrostate: Each neural activity distribution is unique in its overall purpose and represents one unique state of the system. Because of the incredible complexity involved, there is no hope, at least for the moment, to determine the full, detailed neural pattern experimentally and represent it in a mathematically tractable form. However, it is possible to interrogate individual neurons with the implantation of microelectrodes registering the electric spikes of their activity in laboratory animals or in human brains during neurosurgery, and, at the macroscopic level, to register average changes in activity of hundreds of thousands of neurons with noninvasive functional (i.e., time-dependent) tomographic imaging techniques or with electric and magnetic encephalography (Sect. 4.2). Clinical studies of patients with localized brain lesions, later identified in detail in an autopsy, were historically the first methods used to identify the functions of specific brain regions. The time-scale of information-processing operations at the neuronal level is of the order of a few millisecond (see Fig. 4.3 and Sect. 6.3). The time-scale at the meso- and macroscopic levels of interactions between the specific stages (see comments in Sect. 6.3 concerning Fig. 6.2) is of the order of tens to hundreds of milliseconds.

6.1.2 Between any one processing level and the next, there is an information-based interaction in which an input pattern (corresponding to the "sender's" pattern, Sect. 3.5) triggers a response pattern at the next stage which, in absence of any other input, is in univocal correspondence with the input pattern and which has a specific purpose. The corresponding interaction mechanism is determined by the neural "wiring scheme," i.e., the particular anatomic configuration of the synaptic interconnections in the intervening neural network (synaptic distribution and potencies; Sect. 4.1 and Sect. 4.3). There are several "standard" categories of processing mechanisms for specific tasks; some were shown in rudimentary fashion in Fig. 4.1. Among them, *pattern recognition* (elicitation of a unique response whenever a specific pattern is present among a complex or noisy input) is perhaps the most fundamental process operating at the early stages of a sensory system (see Sect. 6.2). Another characteristic process in the visual system is object image rotation, translation and other *topological transformations,* and the (still rather elusive) *binding mechanism* mentioned in Sect. 5.2 required for object recognition. Some of these operations may involve in part "prewired" neural networks with a synaptic architecture that is inherited, i.e., the result of a long evolutionary process (Sect. 4.5). More refined processing mechanisms leading to the actual state of the brain and appropriate response to the current state of the organism and environment involve "plastic" networks, with synapses that have been newly established in learning experiences or whose efficacy has changed during the course of their use (Sect. 4.3).

6.1.3 The short-term or *working memory* is dynamic, represented dynamically in the form of neural activity kept in some "holding pattern," which has a storage lifetime of the order of tens of seconds. The synaptic architecture per se of the information-processing networks represents *long-term* or *structural memory* – both inherited genetic memory (in "prewired" networks) and memory acquired through learning experiences during the lifetime of an organism (Sect. 4.3). A *memory recall* in the neural system consists of the reelicitation or *replay* of that neural activity pattern which represents the recalled information, i.e., is specific to the object, event or concept that is being recalled. As explained in Sect. 4.3, this "holographic" mode of distributed neural memory storage and recall is fundamentally different from the familiar mode of addressed memory used in computers and our daily (non-mental) activities. A very basic brain operation is *associative recall,* in which a key pattern triggers the full pattern representing the recalled information (Sect. 4.3 and Fig. 4.5). The "key" may be just a partial component of the full pattern (see Fig. 4.7), in which case we obtain an *autoassociative recall,* an important operation in sensory perception, as we shall see later.

Associative recall is an example of a "catastrophic" transition of the state of the brain from "not knowing" to "knowing" (even if the knowing is wrong) – it represents the "collapse" of a state of uncertainty with discrete ("categorical") alternatives into one "basis" state of (real or assumed) certitude.[†] That the process of recall is sudden is well-known from our own experience; but it can also be demonstrated numerically with neural network models (Sect. 4.3). Some very convincing quantitative experiments are being conducted with laboratory animals.[‡] In humans, who can trigger memories internally (see Sect. 6.4), "trying to remember something" is a mostly subconscious process in which the brain reviews images related to the sought subject (e.g., events related to a person whose name is being sought) until a "resonance" occurs and the wanted pattern (the name) is triggered. I should point out that the holographic, content-addressed, distributed memory mode of operation (Sect. 4.3) has an enormous evolutionary advantage over a "photographic" (localized) mode of storage and retrieval for two reasons: a) much higher speed of access (faced with an emergency, the brain cannot afford the time it would take to serially review "storage bins" until it finds the wanted information!), and b) a better protection of stored information from physical damage (as clinical studies of brain lesions demonstrate; see also Sect. 4.3). It is thus clear that the brain compensates the inherent slowness of fundamental physicochemical processes (e.g., Fig. 4.3) with the massive parallel-processing needed for its holographic-type operations.

[†] Compare this with the collapse of the state of a quantum system during measurement – but careful: Read the last paragraphs of this section!

[‡] A telling experiment which illustrates the categorical mode of representations in the visual system is the following [40]. A monkey is trained to respond to the picture of a cat or a dog by pressing a corresponding button. Then it is shown the picture of a cat whose features change gradually in computer generated images into those of a dog. While considerable tolerance for change is evident, the transition to the other state of knowledge is very sudden and consistent.

6.1.4 As mentioned in Sect. 4.3, the reelicitation of neural patterns (images) in memory recall requires feedback mechanisms – reverse information-based interactions in which higher level patterns trigger associated patterns at lower levels (see the scheme of Fig. 4.4). A corollary of this is that the expectation or *anticipation* of certain features of a frequently occurring sensory input will trigger neural activity in relevant primary sensory areas of the cortex even *before* an actually occurring feature can elicit the corresponding response; if the expected feature is missing, the corresponding neural pattern appears anyway! This has been verified in many experiments in which the response of featuredetecting neurons in the visual primary cortex is explored by exposing the animal to stimuli in which the expected feature is sometimes present, sometimes missing. Much of sensory perception is indeed based on a process of confirmation and, if necessary, interpolation of missing information or, if the actual input does not match the expectation, correction of the anticipated image. This expectation/interpolation/correction game plays a fundamental role in speech perception, not just at the primary level but at higher levels, too (simplifying the phoneme perception and word recognition) and is also the basis for a theory of the universal aspects of tonal music perception such as consonance/dissonance, tonality modulation, tonal dominance, etc. (for a theory of the affective roles of confirmation and surprise in music, see [94]).

6.1.5 Given the brain's mode of distributed memory and associative recall, incoming information cannot be directed *a priori* to specific processing centers (modules), because the brain does not "initially know" what that information really is. Thus, it must be "broadcast," i.e., spread over many processing regions at the same time, which then may respond or not, according to the particular patterns involved (since addresses play no role in neural information, the brain handles incoming messages as "memos 'To whom it may concern'!"). The corresponding neural patterns are shaped gradually at the different levels as they proceed from the sensory organ through the afferent pathways to the primary cortical receiving areas, and then to the higher centers (see scheme in Fig. 4.4). In other words, *the whole brain* must initially be involved in practically every processing task, until specialized regions can react and engage in parallel processing tasks.[†] We shall discuss this subject in more detail in the next section, but in the meantime I wish to point out the fundamental need for one "grand" cooperative information-processing operation working with "monolithic coherence and synchronism" [71].

[†] The macroscopic synchronization of neural activity revealed in EEG recordings may be an important way for the neural system to ensure appropriate "tuning" of receiving areas to incoming information. But it is doubtful that such global signal can carry very specific information – it may be more alike to the chemical information carriers of the endocrine system, regulating the global states of the brain such as alertness, sleep, etc.

6.1.6 How a specific spatio-temporal neural activity distribution elicited by the sight of an object or by listening to a given sound becomes a specific *mental image* is an old question that has puzzled neuroscientists and philosophers alike. I think that there is a radical answer: The pattern does not "become" anything – the specific distribution *is* the image! There is no "master neuron" or a homunculus: If anything, the homunculus has 10^{11} neurons and is none other than your entire working brain! Let me restate this with an example. When you see a "shiny red apple"; when you close your eyes and imagine a "shiny red apple"; when somebody says the words "shiny red apple"; and when you are reading these lines, there appears a (horribly complex) spatio-temporal distribution of neural activity in many regions of your brain that is common in all cases and represents the cognition of "shiny red apple" and is your mental image – your neural correlate – of a "shiny red apple." It is yours only; physically/physiologically it would be very different from the one that forms in my brain. But the pragmatic *information* it bears would be the same and the activated cerebral regions would be the same!

6.1.7 Venturing for just a moment into a computer analogy, there should be *only one main program* operating at any time – surely, with many subservient parallel subroutines. In other words, we have total parallel processing under tight central management. *Christof Koch* [62] introduces the concept of "neural correlate of consciousness" (NCC) as the minimal set of neurons responsible for a given conscious experience – a "winning coalition" of interacting neurons. The collateral interaction of this "coalition" with other neurons has been termed the "penumbra" of the NCC. However, I prefer to use a terminology more familiar to nonspecialists like "main program" and "subroutines." This mode is Nature's way to prevent competing and eventually contradictory tasks from disrupting life-preserving brain functions and plunging the system into a sort of continuous epileptic state – it is, indeed, a basic expression of *animal consciousness*. In psychiatry there is a syndrome called dissociative identity disorder, in which the patients assume radically different personalities – but never more than one at any given time (e.g., p. 142 in [27]). On the other hand, split-brain patients, whose commissure fibers interconnecting both cerebral hemispheres have been transected for therapeutic reasons, often engage in mutually conflicting actions, but only because each hemisphere, which controls the contralateral side of the musculature and more or less independently processes either language or nonverbal sounds, respectively, cannot communicate with the other. Still, these patients report feeling "like just one person" at all times, even trying to hide or deny any emerging conflict created by the nonconscious hemisphere (and it is the speech hemisphere which controls that personality, e.g. [14]). Finally, we may also consider dreams, in which we always feel as "one" (i.e., the "winning coalition" still works) but some of the subroutines (the "penumbra") fail to engage in coherent fashion.

6.1.8 So far we focused mainly on input. The neural *output* of the brain controls the organism's striate musculature for posture and voluntary or stereotyped movement, some of the smooth muscles of internal organs, the chemical endocrine system, and the general "volume control" by controlling the secretion of brain peptides (Sect. 4.4). An efferent (outgoing) system runs out to the detector organs, antiparallel to the main sensory channels, to exert some limited control of the incoming neural information. The more advanced an animal species, the more options it will have for the response to a given constellation of the environment. This requires *decision-making* based on some *priorities.* Unfortunately, here we tend to think in terms of our own experience with "reasoned" decisions. Animals can neither reason nor engage in long-term planning like humans do (see Sect. 6.4); they follow commands based both on instincts (genetically acquired information) and experience (learned information).

An important part of the subcortical brain, a group of nuclei historically given the umbrella designation of *limbic system,* which we shall examine in greater detail in Sect. 6.4, acts as a sort of "police station" checking on the state of the environment and the organism, directing the animal's *attention* and *motivation,* and making sure that the output – the *behavioral response* – is beneficial to the survival of the organism and the propagation of the species based on evolutionary and ontogenetic experience. The limbic system works in a curious "binary" way by dispensing feelings of "reward or punishment" such as: hope or anxiety, boldness or fear, love or rage, satisfaction or disappointment, happiness or sadness, etc. These are *emotional states* which have the mission to evoke the *anticipation* of pleasure or pain whenever certain environmental events are expected to lead to something favorable or detrimental to the organism, respectively. Since such anticipation comes *before* any actual benefit or harm could arise, the affective mode helps guide the organism's response into a direction of maximum chance of survival, dictated by information acquired during both evolution and experience. In short, the limbic system directs a brain to *want* to survive and find the best way of doing so in real-time, given unforeseen circumstances that cannot be confronted on the basis of preprogrammed (instinctive, automatic) instructions alone. Using a somewhat slippery term difficult to define precisely, the limbic system attaches subjective *meaning* to the percepts and the alternatives for pertinent behavioral reactions. Obviously, to work, the limbic system must interact with the highest levels of brain processing, where input is interpreted and output is planned. Although in item 6.1.7 above we have associated animal consciousness with the existence of a "main program" (or "winning coalition" of participating neurons), it is really the coherent interplay between cortical and limbic functions which defines *core consciousness* ([27], see also Sect. 6.4).

6.1.9 All the above goes for higher animal brains in general. The *human brain* has extra capabilities, some of which were already mentioned briefly in Sect. 5.1. We shall dedicate an entire section to them (Sect. 6.4); here we shall only mention self-consciousness, the human capability of making *one representation of all representations* and the associated feeling of being totally in control. This higher-order level representation in the human brain has cognizance of core consciousness and can manipulate independently this "main program" or "winning coalition" of neurons. It can construct an image of the *act* of forming first-order images of environment and organism, as well as of the reactions to them. It can overrule the dictates of the limbic system, but it can also stimulate limbic response without any concurrent external or somatic input.

Finally, let me briefly comment on *what the brain is not*. It is not the equivalent of a computer: It does not operate with digital gates (despite of the fact that APs are sometimes viewed as digital signals) and it does not have any software (Sect. 4.2). Whereas it is a super-parallel-processing system which follows genetic instructions (embedded in the wiring scheme of mainly the limbic structures), by and large it makes its own rules as it accumulates information on the environment in real time – indeed, the brain is the quintessential adaptive, self-organizing system. To imply, as it is frequently done, including by myself (see item 6.1.7 above!), that the brain works like a computer is really an insult to both. It is an insult to the brain because computers are so primitive when it comes to processing complex analog *pragmatic information* (Sect. 3.5 and Sect. 3.6). It is an insult to a computer because the brain is so slow in handling digital *Shannon information* (Sect. 1.3).

The brain is not the equivalent of a quantum computer, either. It does not operate on the basis of quantum mechanisms that transcend into the macroscopic domain, although some distinguished scientists think so. Indeed, *Penrose* [85] proposed that consciousness may be a quantum process, given the quantum-like behavior of some brain functions – see paragraph 6.1.3 above, where I have deliberately chosen language concerning the "collapse" of a brain state to make it sound like the collapse of a quantum wave function (Sect. 2.8 and Sect. 5.6). Microtubules made of a protein called tubulin, responsible for the skeletal structure of all cells, have been invoked as possible units of quantum behavior. These proteins, like the motile proteins of the actin family in muscle fibers, have two possible states (extended and contracted) and it has been speculated that in the neurons they may exhibit the behavior of Qbits (be in a *superposed* state, Sect. 2.7), making the brain a phenomenal parallel-processing quantum computer. However, is has been shown [110] that because of decoherence (Sect. 5.6), such a macroscopic quantum system, working at body temperature, could never maintain superposed states for more than a tiny fraction of time (10^{-13} s to 10^{-20} s). In my view, a quantum hypothesis for the brain, while intellectually attractive, is not necessary: Properties of cooperative processing, sudden transitions, collapse, associative recall, etc., appear in rudimentary fashion with all their characteristics in (ultra-classical) numerical neural network modeling, as discussed in Sect. 4.2. Certainly, there is no equivalent of consciousness in those models, but this may be "just" a matter of the number of cooperatively interacting elements – maybe after $\sim 10^{10}$ mutually interacting information elements there is a "magic threshold" for the appearance of consciousness?!

There are other things the brain does not do. It does not store information in some individual "memory molecules" as some scientists have believed for a long time. Likewise, it does not store integral (categorical) information in individual neurons, although current parlance contributes to this erroneous image: One speaks of "grandmother neurons" which respond with increased firing rate when a picture or the memory of one's grandmother comes up (see Sect. 6.3). Neurons with such properties do in effect exist, especially in the temporal and prefrontal lobes, but they are among many others[†] that participate in the specific spatio-temporal pattern that is the neural correlate of what happens when we see, hear or think of our grandmother (see in particular Sect. 4.2 and point 6.1.3a above). In general, individual neurons usually participate in many different information-processing tasks.

The next sections deal with some details of neural hardware; as I did in Chap. 2, I suggest to readers less interested in this special subject *to jump directly into Sect. 6.4*. But again, as I warned on the previous occasion, they may miss some important points – this time concerning information processing in the brain.

6.2 Sensory Input to the Brain: Breaking Up Information into Pieces

Environmental representation and prediction are fundamental operations of the brain. The vertebrate brain must construct accurate "floor plans" of the relevant surroundings (e.g., a dog knowing how to find a given room in a house), and it must establish accurate "time tables" for the course of relevant events (a dog anticipating a meal on the basis of some learned temporal ritual). To accomplish this, the brain must discover the spatial and temporal configurations of objects that are relevant to the organism, and the cause-and-effect relationships among them. Such correlations, and the symmetries and periodicities that the brain discovers during its interaction with the environment, play an essential role in all cognitive tasks, helping to sort out redundancies from an unwieldy sensorial input. In this section we examine some critical stages of sensory information processing in the two most important and sophisticated sensory systems: visual and auditory.

Human beings (and higher vertebrates) are visual animals: A large portion of cortical tissue (the occipital lobes in the back of your head) is dedicated al-

---

[†] A quote from *von der Marlsburg* [71], p. 203 comes in handy: "... it does not help staring at the neuron whose signal is linked with some elementary sensation – the essence of that sensation is the reaction patterns of the whole brain."

most exclusively to the reception and initial processing of visual information. Much of the relevant information is *spatial*, pertaining to the two-dimensional patterns optically (i.e., physically) displayed on the retina. This is the visual sensory organ proper that holds primary detector cells, about 100 million "rod" cells (sensitive to dim light) and the 5 million "cone" cells (for bright light), distributed with uneven density (maximum in the fovea, where the direction of gaze is projected). A lot of primary information processing such as contrast detection and motion detection is already performed in the neural network of the retina (sometimes called a "spilled-over" part of the brain) – to the point that the incoming information can be compressed (Sect. 1.4) into just one million ganglion cells whose axons form the optical nerve of the corresponding eye. The fact that we have two eyes with identical afferent processing channels allows us to form a stereoscopic 3D picture out of both 2D physical images on the retinas (see below). Time plays a role in the visual system, but only on a longer-term time-scale of hundreds of milliseconds (to detect motion).

The situation is quite different in the auditory system, which has an eminently *temporal* function, detecting the vibration pattern of the essentially one-dimensional basilar membrane, a band of varying elasticity stretched about 35 mm along the snail-like cochlear duct in the temporal bone. The basilar membrane holds the organ of Corti, consisting of four rows of a total of about 16 000 hair cells. These detector cells have tiny cilia which, if bent by the surrounding cochlear fluid, generate electric signals that the hair cells pass on to spiral ganglion cells, whose axons form the acoustic nerve. Any incoming sound is converted into cochlear fluid oscillations which, in turn, cause the basilar membrane to vibrate like a waving flag; since the elasticity of the membrane decreases by a factor of 10 000 from its beginning to its end point, the membrane frequency response (resonance) varies along its length. As a result, the vibration pattern of the basilar membrane is an approximate (self-consistently bad!) Fourier transform of the incoming temporal pattern of acoustical pressure oscillations. Hair cells respond in a very nonlinear way to either the velocity or the displacement of the basilar membrane (see Fig. 2.22 in [94]), encoding sound frequency both in spatial form (position of resonance region) and time ("bunching" of neural impulses). There are only 32 000 or so fibers in the human acoustic nerve (compared to a million in the optic nerve), but this does not mean that the acoustic system is less sophisticated; what happens is that it works in only one dimension (the transformed frequency spectrum of the input) but is able to resolve time differences that are smaller than a single neuron's time resolution!

Please note that in the two preceding paragraphs we have only referred to one kind of information, encoding location in space (directional distribution of incoming photons) in the visual system, and location in frequency space (frequency spectrum of incoming sound waves). Intensity (photon flux, acoustic power flow) provides another fundamental kind of information, encoded at the afferent levels in the form of number and type of responding fibers and their firing rates. Color and timbre are sensory attributes extracted through integration of several characteristics of the afferent information and formed at the cortical, probably postprimary, level. Quite generally, in the optic and acoustic systems four fundamental levels of information-processing can be identified after the transduction of the physical signals into neural patterns in the sensory organs.

1. The afferent pathways, where the initial neural patterns are transformed but still kept more or less in topological conformity with the physical input distribution (retinal "pixels" defined by elements of solid angle in vision, the critical band defined by elements in frequency space in audition).
2. The primary cortical receiving areas (the striate cortex in the occipital lobe for vision, Heschl's gyrus in the temporal lobe for hearing).
3. The so-called association areas and other regions of the cortex where the origin of the signals is identified.
4. Integration of the identified incoming information into the "whole picture" of conscious perception in the prefrontal lobes. All this is being constantly watched, encouraged or censored by "Big Brother" sitting in the subcortical structures: the almighty limbic system (Sect. 6.1 and Sect. 6.4).

Figure 6.1a,b show in flowchart form the afferent input channels of the optic and acoustic systems, as they convey neural information from the sensory organs through the brainstem and subcortical regions to the primary sensory cortices. We show them here only to point out some key stages of environmental information processing before the information reaches the cortex, as well as the transformations of purpose and meaning it undergoes. Anatomical exploration of these channels was done in the past with staining techniques in brains of cadavers (I drew these diagrams for my lectures nearly 30 years ago based on such information from neuroanatomy books); now it is possible to explore the functional aspects in living animals with microelectrode recordings, and in human subjects with the noninvasive techniques of functional MRI (see e.g., [81] and Sect. 4.2). This latter task is not easy and is still at an early stage as a tool for basic research.[†] We said above that there is an important first information processing stage in the retina. A somewhat equivalent stage in the acoustic system is the spiral ganglion in the cochlea and the cochlear nucleus located in the brainstem. Notice by comparing both

---

[†] Since what is being imaged in an fMRI is change in blood oxygenation, corrections for real-time cardiac pulsations are necessary (each image has to be taken at a specific phase of the heart cycle). For the auditory channel the situation is complicated by the substantial and unavoidable noise of the imaging equipment (cooling system, magnetostriction noise) which interferes with what is being measured.

charts that on-the-way processing and cross-channel communication are more apparent in the acoustic system. This is only natural because time is of prime importance here; for instance, it is believed that in the medial superior olive a neural crosscorrelation mechanism determines interaural timing differences for sound localization to an accuracy of 20 µs (a similar mechanism in bats is as much as ten times more accurate for their sonar echolocation system).

In the optic system on the other hand, there are no interactions between the right and left channels (careful: "right" and "left" refers here to channels from the right and left *visual fields,* respectively, not right and left eyes – look at the sketch in Fig. 6.1a!). Indeed, information from the right (left) half-space, projected onto *both* left (right) hemi-retinas, is conveyed to the *left* (right) primary cortical area, so that information on events in one visual hemisphere go to that cerebral hemisphere which controls the corresponding musculature (the left hemisphere controls the right side of the body and vice versa). A person with a massive lesion in, say, the left primary visual cortex has a blocked-out right visual field in *both* eyes; yet a person with a similar lesion in the left primary auditory cortex can hear sounds fed monaurally into the right ear. Notice the curious arrangement of the lateral geniculate nucleus in the visual system, and the two channels appertaining to the right and left hemi-retinas: they convey information to neighboring but separate stripes in the input layer of the primary visual cortex (the V1 region, [62]) corresponding to either the contralateral or ipsilateral hemi-retinas (but always the same visual half-space). At deeper layers of this part of the cortex the interaction between the corresponding columns in the cortical tissue yields information on the distance of an object from the ocular plane.[†]

In the acoustic system, there is no such convergence and intersection of bilateral neural information at the primary cortical level; as explained above, the left–right intercomparison most likely occurs at a very early stage in the medial superior olive to avoid decoherence, i.e., a loss of the binaural information on acoustical signal time difference (intraaural phase difference). Instead, in the acoustic system there is a pronounced *right–left functional asymmetry* at the cortical level: The left temporal lobe (in 97% of all individuals) specializes in *serial* and *analytic* acoustic information processing such as required for language, whereas the right lobe deals mainly with *holistic* and *synthetic* processing, which for instance is required in music perception [94]. This division of tasks is not limited to the acoustic system – to a certain extent it also

---

[†] If you cover your left eye and see (with the right eye) a nearby object located in the *right* visual field, your brain forms an image of the object using information processed in the contralateral stripes (see Fig. 6.1a) of the left visual cortex. If you now switch the cover to the right eye, you will see the image displaced horizontally; that image, seen from a slightly different optical perspective, is now formed in the other (ipsilateral) stripes (still in the left visual cortex) receiving information from the left eye. If you now look with both eyes, you will notice what is called binocular fusion: Information is now extracted from both types of stripes and you obtain one single image with a sense of distance and 3D form.

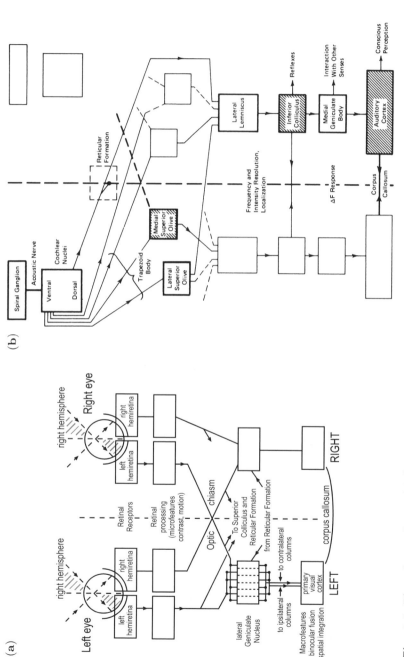

Fig. 6.1. Flow charts of neural information in the visual and auditory afferent channels between the sensory organ and the primary cortical receiving area, depicting principal transmission channels, processing stations (nuclei) and routes to other parts of the brain (not in scale!). (**a**) Visual afferent system (from *Roederer* [90]). (**b**) Auditory system (from *Roederer* [94])

applies to vision (analysis of time-sequences mainly in the left hemisphere, analysis of spatial configurations preferentially in the right hemisphere [17]). There even is evidence that cerebral lateralization exists in some animals for which the distinction between serial and integral processing is crucial for survival.

It is important to note that some information from the afferent optic and acoustic pathways converges at the superior colliculus (not shown), which means that there is intrasensory mixing already at a subcortical level. This may contribute to synesthetic effects, such as visual illusions while hearing certain sounds (e.g., seeing colors while listening to music), or acoustic illusions while seeing luminous images (e.g., hearing sounds while watching the polar aurora). Finally, note the connections to the so-called reticular formation (like the limbic system, an umbrella name), a veritable neural "switch-board" consisting of a network of small nuclei in the back of the brainstem which receives information from, and delivers it to, all sensory systems, the body and higher brain structures [27] (Sect. 6.4). The reticular formation is responsible for activating or inhibiting information processing, and controls wakefulness and sleep, awareness, even consciousness (damage to components of this diffuse network, as happens in a sharp blow to the back of the neck, can lead to irreversible coma). Not shown in Fig. 6.1b is a network of efferent fibers, which carries information from the upper stages to the lower ones and terminates in the cochlea. The peripheral part of this efferent system, called the olivocochlear bundle, plays an important role in the control of incoming acoustic information (although there are only about 1600 efferent fibers reaching each cochlea).

Let me point out some general features of the acoustic system from the informational point of view. At the initial stages of the acoustic system (cochlear nucleus) there is a very specific geometric correspondence between activated neural fibers and the spatial position of the source stimulus (resonance regions) on the basilar membrane. As one moves up, however, this correspondence is gradually lost (except in an anesthetized state). The number of participating neurons increases dramatically and the neural response becomes increasingly representative of complex features of the sound signal, being more and more influenced by information from higher levels on the behavioral state and the performance of the individual. This is borne out clearly in the fMRI responses at different levels of the auditory pathway: Measuring the percent signal changes of neural activity when a 30 s noise burst is fed into an ear [75], the activity is nearly constant during the noise stimulus in the inferior colliculus (see Fig. 6.1b), with specific rise and decay slopes; higher up in the medial geniculate body there is already a marked change during the (constant) noise period, and at the cortical level, the responses are clearly onset ("here come the noise!") and offset ("it's over!") signals. The change in pattern (for the same input) clearly indicates a changing purpose of the corresponding information-driven interaction mechanisms (Sect. 3.5). On the other hand, psychoacoustic experiments show that information on the phase of a vibrational pattern (i.e., its timing) is preserved at least until the right and left channels have been interconnected for the first time (in the trapezoid body): One can detect beats between two mistuned tones fed separately into each ear, even when there was no chance for a physical superposition of both signals (e.g., [94]). Contralateral channels are "better" information carriers than ipsilateral (same-side) ones: If conflicting information is presented separately into both ears, the contralateral channel tends to override the information that is carried by the ipsilateral channel (it is more difficult to understand speech if it is fed to the left ear while you are fed another signal, equally loud, into the right ear, than vice versa – provided your speech centers are on the left side). In the optic afferent system, the geometric correspondence with signal position on the retina is preserved through the stages; what changes during the course (starting in the retina itself) is the representation of optical features (from intensity gradients and other "smooth" 2D features projected on the retina to contrast-enhanced features, motion signals, then lines of different directions, etc. in the primary cortex).

6.3 Information Integration: Putting the Pieces Together Again – in a Different Way

Neurons in the primary cortical areas to which the afferent transmission system from a sensory organ is wired, are "feature detectors." If I look at a tree, there is no activated region on the cortex that has the form of a tree. If I hear a trumpet sound, nothing blips in my cortex with a pattern that emulates the acoustic vibration patterns of a trumpet. The afferent processing stages and the primary cortex have taken the original images apart, and remapped their component features in quite different ways and locations. In the visual cortex, while individual neurons do have a specific receptive field, that is, a small solid angle fixed to the eyeball for incident photons to which they respond, they only do so for certain well-defined patterns appearing in that receptive field, such as a dark or light bar inclined with a specific angle, an edge moving in a certain direction, and so on (e.g., [62, 72]). In the primary auditory cortex there are neurons that are "tuned" to a specific frequency interval, but they tend to respond only to certain complex sound stimuli in that frequency domain and specific transients (the famous "meow detectors" in cats). Now these disjoint patterns have to be bound together, in such a way that features belonging to one and the same object (spatial or temporal) elicit a pattern that is specific and univocal to *that* object – regardless where in visual space or frequency space it is located (remember that "shiny red apple" in point 6.1.6!). In other words, the incoming information has to be assigned into categories that have to do with *meaning*. In vision, responses to edges and lines belonging to the same object have to be transformed into *one* neural pattern that is in one-to-one correspondence with that object; in hearing the spatially dispersed responses corresponding to the resonance regions of harmonics of a musical tone have to be transformed into *one* pattern specific to the pitch and timbre of a single musical tone. At the linguistic level, the complex and changing patterns elicited by speech must be transformed into patterns that correspond to specific phonemes, words and sentences.

Let us examine in more detail how the various levels or stages of information processing and representation operate in the brain from the primary area upwards. For that purpose, we examine the sketch of Fig. 6.2 depicting an oversimplified view of the visual system. As described in the previous section, the neural circuitry in the periphery and afferent pathways up to and including the so-called primary sensory receiving area of the cortex (stage 1 in the figure) carry out some basic preprocessing operations mostly related to *feature detection.* The next cortical stage 2a carries out the above mentioned *feature integration* or *binding* process, needed to sort out from an incredibly complex input those features that belong to one and the same spatial or temporal object. A second stream (stage 2b) executes geometric transformations that assign "identity" to information coming from the same 3D spatial object seen at different distances, positions and orientations. In other words, both operations transform radically different patterns (originating in the different projections on the retina from one object) into single patterns that are in correspondence with the topological properties of the form of that object (for object recognition 2a), and with the spatial position of the object in the environment (for eventual motor actions 2b). Some of these transformations may be learned, i.e., the result of experience during the first months of an infant, with the sense of touch providing a signal of "uniqueness" to an object that is being handled and simultaneously looked at (it may be not by chance that region 2b is near the somatosensory area). They certainly can be learned at a later stage in life (for instance one can learn to read upside-down or mirror-image text).

Ablation studies with animals have shown that at this stage the brain "knows" that it is dealing with an object, but it does not yet know *what* the object is. This requires a complex process of comparison with existing, previously acquired information (stage 3 in the figure) and must rely on the process of associative recall. For instance, in the human medial temporal lobe neurons were found that respond whenever faces of certain persons, environmental scenes, specific objects or animals are seen [72]. As one moves up along the stages of Fig. 6.2, the information processing becomes less automatic and more and more centrally controlled; in particular, more *motivation-controlled* actions and decisions are necessary, and increasingly the *previously stored* (learned) information will influence the outcome. As to the time scale, single cell recordings in *monkeys* show the following progression of activation times or "latency" counted from the detection of a flashed stimulus at the retina (reported in [111]); stage 1 in Fig. 6.2: 40 ms to 60 ms; stage 2b: 80 ms to 100 ms; stage 4: 100 ms to 130 ms. The quoted times are about 30% shorter

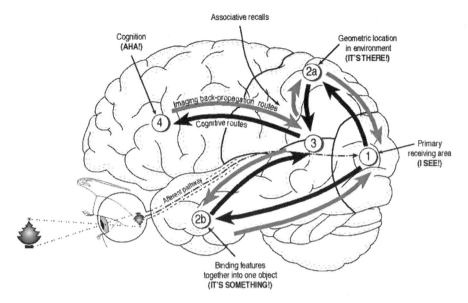

Fig. 6.2. Ascending information routes and processing levels for visual information (from *Roederer* [96]). For the timing (latency) of activity from detection to cognition, see text. The routing through lower, subcortical levels (not shown) checks on current subjective relevance of the information on its way to the prefrontal cortex. Through feedback pathways, the imagination of a given object triggers neural activity distributions at lower levels that would occur if the object was actually perceived by the eye

than those reported for visual processing in the human brain, possibly related to the smaller-size brains of primates.

The acoustic system has some equivalent processing stages (from the informational point of view), except for the existence of the striking hemispheric lateralization, i.e., the above mentioned division of sequential vs. synthetic tasks (e.g., [17]). The evolutionary reason for this division is probably the brain's own version of "time is money": spoken language processing, generally accepted as the most distinguishing ability of human information processing (see next section) and perhaps the most significant step in human evolution (see Sect. 6.4), puts enormous demands on the rate and speed of cerebral information processing. The brain simply cannot afford the 50 or more millisecond it takes to exchange information between both hemispheres when it comes to speech perception. So the fast sequential tasks are kept together in spatial proximity in one temporal lobe, which happens to be the left one in 97% of the individuals. We have a similar situation with present-day electronic computers: The main limitation to their speed is given quite simply by the spatial distance between computing units! This concentration of specialized activity in one temporal lobe freed a lot of neural circuitry in the opposite temporal lobe for other, less time-demanding synthetic acoustic tasks such as the integration in frequency space required for timbre perception, harmony, melodic lines and, at higher levels in the parietal and frontal lobes, musical instrument recognition and holistic music perception, as well as, for nonmusical sound, recognition of the speaker, and general synthesis and integration into the acoustic scene.

In the speech hemisphere (also called the "dominant" hemisphere), there is a well-delineated three-step processing path [13] from the superior temporal gyrus (next to Heschl's gyrus, the primary receiving area), to the phonemic pattern recognition systems around the superior temporal sulcus, and the first stage of lexical–semantic processing in the ventro-lateral part of the temporal lobe.[†] From there, the information proceeds to several "higher" areas, including the posterior cingulate (the cingulate gyrus is an important part of the cortex, buried deep in the middle groove, that interacts two-way with many other cortical and subcortical areas), the prefrontal cortex and angular gyrus, in a very complex and not yet fully explored series of steps for the full linguistic analysis, in which, like in the visual pathways, associative recall mechanisms play a fundamental role. As we shall see in the next section, there is an intimate connection of these processing stations with those responsible for human thinking. The acoustic information processing in the minor hemisphere is more diffuse and less explored (research in music processing has a lower priority and receives less funding than speech!). Basically, the equivalents to phoneme-lexical processing would be complex tone, chord and melodic analysis (for a more detailed discussion of sound lateralization, see Sect. 5.7 of [94]).

We mentioned in Sect. 4.3 and again in point 6.1.3 that a memory recall consists of the reelicitation of a neural pattern that is in univocal correspondence with what is being recalled. There is more to it. The paths sketched in Fig. 6.2 can also operate *in reverse:* There is now ample experimental evidence that the memory recall (or, in humans, also the imagination) of a given object, triggers neural activity distributions at the various lower levels that would occur if the object was actually seen (see also point 6.1.3 and comments on the "shiny red apple" in item 6.1.6; for details of this "top down control" see also [79]). In the auditory system, "internal hearing" operates on the following basis: The imagination of a melody or the recall of spoken words is the result of the activation, or "replay," of neural activity triggered somewhere in the prefrontal cortical areas (depending on the specific recall process), which then feed information back down the line of the auditory processing stages creating sensory images without any sound whatsoever entering our ears. In animals such backward transfer of information happens "automatically," triggered by associations or other environmental input; in

---

[†] For the location of the cerebral regions mentioned in this section, see the sketches in [27, 62].

humans it can be willed without an external input (see next section). The experimental evidence does not yet prove that in such a backward-propagating situation the elicited spatio-temporal distribution of activity is actually identical to that which would occur for an equivalent external (sensory) input, but at least it shows that the participating areas are the same. When the experiments involve thinking about words (e.g., [86]), i.e., semantic processing during absence of acoustic input or during silent reading or lip-reading (in which case the incoming information is visual but the tested areas are acoustic), fMRI imaging shows the active involvement of four regions: the angular gyrus, dorsal prefrontal cortex, posterior cingulate and ventral temporal lobe, listed here in reverse order from the above mentioned case of incoming speech information processing.

6.4 Feelings, Consciousness, Self-Consciousness and Information

What remains to be discussed is the question of *motivation* (to know or to imagine) and the *value* or meaning of something being known or imagined (see Table 1.1). For that purpose we turn again to Fig. 6.2 and to the fact that from all the stages shown there are also interactive routes to lower, subcortical nuclei. As mentioned in Sect. 6.1, one of the phylogenetically old parts of the vertebrate brain is the so-called *limbic system,* which works on the basis of genetic information represented in mostly prewired (inherited) networks. One uses this term as a short-hand for several deep subcortical structures, although it should be mentioned at once that brain scientists are very reluctant to use "umbrella" designations for processing centers that interact in different ways depending on the actual context of the task in question [27]. The structures in question integrate a group of subcortical nuclei located near the midline of the brain in the brainstem and the thalamus, which in conjunction with the hypothalamus (a control center of the chemical information system, Sect. 4.4), the basal forebrain and the cingulate cortex perform the following tasks: 1. construct neural representations of the state of the organism; 2. selectively direct information storage according to the relevance or *meaning* for the organism; 3. regulate body functions; and 4. mobilize motor output (behavior). This is schematically summarized in Fig. 6.3; for a recent review see [33].

Emotion (controlled by the deeper structures) and motivation and will (controlled by the anterior cingulate cortex) are integral manifestations of the limbic system's function. This system constantly challenges the brain to find alternatives and pertinent solutions, to probe the environment, to overcome aversion and difficulty in the desire to achieve a goal, and to perform certain actions even if not needed by the organism at that moment (like animal curiosity and play). In all its tasks, the limbic system communicates interactively with the cortex, particularly the prefrontal regions, relating everything

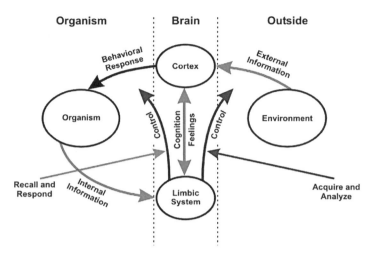

Fig. 6.3. Basic functions of brain regions collectively known as the limbic system – the brain's "police station" that assures that cognitive functions and behavioral response are beneficial for the organism and the propagation of the species. The limbic system controls emotion and feelings and communicates interactively with higher processing levels of the cortex, particularly the prefrontal regions, relating everything the brain perceives and plans to the needs of the organism and vice versa. In higher species, the coherent operational mode of the cortico-limbic interplay gives rise to consciousness. The human brain is able to map and coordinate this interplay at an even higher level, as well as to overrule limbic dictates; this leads to self-consciousness. After *Roederer* [94]

the brain perceives and plans to the state of the organism and vice versa (Fig. 6.3). In short, the aim of this system is to ensure a behavioral response that is most beneficial to the organism and the propagation of the species according to genetically acquired information – the so called *instincts* and *drives*.

As mentioned already in Sect. 6.1, to carry out its functions, the limbic system works in a curious way by dispensing sensations of reward or punishment: hope or anxiety, boldness or fear, love or rage, satisfaction or disappointment, happiness or sadness, etc. Of course, only human beings can report to each other on these *feelings*, but we have every reason to believe that higher vertebrates also experience such a "digital" repertoire. We mentioned briefly in item 6.1.8, the possible evolutionary advantage of this mode of operation. Still, I feel that it is one of the greatest biological mysteries – perhaps even more so than the cooperative mode of operation that gives rise to consciousness and mental singleness. Why does an open wound hurt? Why do we feel pleasure eating chocolate, scratching a mosquito bite or having sex? How would we program similar reactions into a robot? Of course, the designer could program a robot to emit crying sounds whenever it loses a part and to look for a replacement, reach out toward loose screws and tighten them, seek an electrical outlet whenever its batteries are running low, join a robotess to construct little robotlets, etc. This would merely be programming emotion-related behavioral output, but how would we make the robot actually *feel* fear or *want* pleasure, before pain or pleasure actually occur? Plants cannot respond quickly and plants do not exhibit emotions; their defenses (spines, poisons) or insect-attracting charms (colors, scents) have been programmed during the slow process of evolution (Sect. 4.5). Reactions of nonvertebrate animals are neural-controlled but "automatic" – there is no interplay between two distinct neural information processing systems and there are no feelings controlling behavioral response. I believe that without the guiding mechanism of a limbic "control station" intelligence could not have evolved [98].

Cortical activity is consistently monitored by the limbic structures and so is the information on the state of the organism (Fig. 6.3); the resulting information-based interactions in turn are mapped onto certain cortical regions, especially in the prefrontal areas. In this interplay a balance *must* be achieved, or the organism would succumb to conflicting behavioral instructions. Somehow out of this balance or compromise, based on real-time input, memory of experienced events and instinct, emerges one primary goal for action at a time – this is the essence of animal *core consciousness* [27] and mental singleness. Note that there are specific periods of time involved: 1. information from the past (both the distant past built up over a long time in Darwinian evolution collectively representing the instincts (Sect. 4.5), and the individual's past as it has impacted the organism[†]); 2. information from the present on the state of the body and the state of the environment; and 3. a narrow window of time for short-term predictions of perhaps a few tens of seconds[‡] (probably defined by the capacity of the short-term memory (Sect. 4.3)). It is concerning this last window of time that the most striking differences between infra-human and human brains are found.

From the neurophysiological and neuroanatomical points of view the human brain is not particularly different from that of a chimpanzee. It does have a cortex with more neurons and some of the cortico–cortical fascicles have more fibers, but this difference is of barely a factor of 2 or 3. More significant is the total number of synapses in the adult human brain. Is the difference in information processing capabilities only one of quantity but not one of substance?

Aristotle already recognized that "animals have memory and are able of instruction, but no other animal except man can recall the past at will." More specifically, the most fundamentally distinct operation that the human,

---

[†] The memory of events that have had an impact on the organism is called the *autobiographical memory* [27, 62]. For instance, knowing the laws of physics is not part of the autobiographical memory, but understanding them is.

[‡] There is no evidence that, for instance, an infra-human predator can decide *today* what strategy it will be using tomorrow.

and only the human, brain can perform is to recall stored information as images or representations, manipulate them, and restore modified or amended versions thereof *without any concurrent external sensory input* [91]. In other words, the human brain has internal control over feedback information flow as depicted in Fig. 6.2; an animal can anticipate some event on a short-term basis (seconds), but only in the context of some real-time somatic and/or sensory input (i.e., triggered by "automatic" associative recall processes (Sect. 4.3)). The act of information recall, alteration and restorage *without* any external input represents the *human thinking process* or *reasoning* [91]. *Young* [114] stated this in the following terms:

> Humans have capacity to rearrange the "facts" that have been learned so as to show their relations and relevance to many aspects of events in the world with which they seem at first to have no connection.

And *Bickerton* [12] writes:

> . . . only humans can assemble fragments of information to form a pattern that they can later act upon without having to wait on . . . experience.

A brief interlude is in order. Categorical statements like those in the preceding paragraph, which trace a sharp boundary, or "ontological discontinuity," between human and animal brain capabilities, are disputed by animal psychologists and many behavioral scientists. They point out the fact that apes (e.g., chimpanzees) and even some birds like corvids (e.g., the familiar Alaskan ravens circling my house in $-40\,^{\circ}$C weather) use tools, construct nests, assemble food caches and exhibit a highly sophisticated social behavior, thus demonstrating ability of complex cognition and mental representation of time (e.g., [36]). However, these behavioral activities are not based on the knowledge of *how* a tool works or *why* this or that feature of a nest is better suited for present needs – just as a parrot which has learned to imitate human speech sounds does not know *what* he is saying (sorry, parrot owners!). Animal behavior like tool-making and shelter-building that looks very sophisticated to us (and indeed is!) follows "blueprints" which differ substantially from the "blueprints" followed by a human being who is building a hut or designing a mansion. The latter are conceived "on the spur of the moment" based on information "from the future" – i.e., as a result of long-term planning (see below) – and they can be modified radically in real time not only to take into account unforeseen circumstances but because of changing ideas, i.e., changing mental images (models!) of the goal to be achieved. Animal blueprints, instead, are handed down by the process of evolution. It is, however, important to take into consideration what we have stated in Sect. 4.5: Not everything that is the result of DNA-triggered action has to be reflected in information contained in the genome. Nature masterfully takes advantage of the properties of things "out there" and what they can do because of their intrinsic complexity or self-organization.

The capability of recalling information without any concurrent input had vast consequences for human evolution. In particular, the capability of reexamining, rearranging and altering images led to the discovery of previously overlooked cause-and-effect relationships – this is equivalent to the creation of new pragmatic information and reduction of algorithmic information (see Sect. 1.4). It also led to a quantitative concept of elapsed time and to the awareness of future time. Along this came the possibility of *long-term prediction* and *planning* ("information about the future") [95, 107], i.e., the mental representation of things or events that have not yet occurred (again, this should not be confused with the capacity of higher vertebrates to anticipate the course of current events on a short-term basis of tens of seconds). Concomitantly with this came the postponement of behavioral goals and, more generally, the capacity *to overrule the dictates of the limbic system* (think of sticking to a diet even if you are hungry) and also *to willfully stimulate the limbic system,* without external input (e.g., getting enraged by thinking about a certain political leader). In short, the body started serving the brain instead of the other way around! In all this, the anterior cingulate cortex may play a fundamental role of transforming intention into action. Mental images and emotional feelings can thus be created that have no relationship with momentary sensory input – the human brain can go "off-line" [12]. Abstract thinking and artistic creativity began; the capacity to predict also brought the development of beliefs (unverifiable long-term predictions), values (priorities set by society) and, much later, science (improving the capacity to predict).

In parallel with this development came the ability to encode complex mental images into simple acoustic signals and the emergence of *human language.* This was of such decisive importance for the development of human intelligence that certain parts of the auditory and motor cortices began to specialize in verbal image coding and decoding, and the human thinking process began to be influenced and sometimes controlled by the language networks (e.g., [86]) (this does not mean that we always think in words!). Finally, though only much later in human evolution, there came the deliberate storage of information in the environment; this *externalization of memory* led to the documentation of events and feelings through visual symbols and written language, music scores, visual artistic expression, and science – to *human culture* as such. And it was only very recently that human beings started creating artifacts capable of processing information and entertaining information-based interactions with the environment, such as servomechanisms, computers and robots, and accelerating the old genetic modification process of animal breeding with genetic engineering and cloning.

It is important to point out that the capabilities of recalling and rearranging stored information without external input, making long-term predictions, planning and having the concept of future time, stimulating or overruling limbic drives, and developing language, most likely all co-evolved as one sin-

gle neural expression of human intelligence. At the root of this development
from the informational point of view lies the human capability of making
one representation of all representations. There is a higher-order level of representation in the human brain which has cognizance of consciousness and
which can manipulate independently the primary neural representation of
current brain activity both in terms of cognitive acts and feelings. It can construct an image of the *act* of forming first-order images of environment and
organism, as well as of the reactions to them. In other words, the informational processes schematically depicted in Fig. 6.3 have a representation at a
higher level in humans, and as happens with the routes of Fig. 6.2, a retro-
propagation allows those higher-level patterns to influence neural patterns at
the lower levels. There is no need to assume the existence of a separate neural
network; there is enough information-handling capacity in the cortical and
subcortical networks that can be shared with the processing of lower-level
representations (except, perhaps, that there may be a need for a greater involvement of the prefrontal cortex and the possibility that language networks
may participate in important ways).[†]

The capacity of retrieving and manipulating information stored in memory without any external or somatic trigger; the feeling of being able to
observe and control one's own brain function; the feeling of "being just one"
(even in the most severe case of multiple personality disorder as mentioned in
item. 6.1.7); the capacity of making decisions that do not depend on real-time
environmental necessity and somatic input; and the possibility of either overruling or independently stimulating the limbic dictates, collectively lead to
what we call *human self-consciousness* (close but not equal to the "extended
consciousness" defined by *Damasio* [27]). It seems to me that self-consciousness is far more than just a feeling – it represents the capability of some very
unique information processing *actions*. A useful metaphor for this would be
the following: While consciousness is "watching the movie that is running in
the brain," self-consciousness is the capacity of human brains "to splice and
edit that movie and even to replace it with another one" [95].[‡]

At this point we seem to have lost all direct contact with the real purpose
of this chapter, announced in its first paragraph. But this is not so: I hope
that I have indeed laid the ground for a better understanding of the neurophysiological underpinnings of "knowing," "imagining," "expecting" and

---

[†] Injecting a barbiturate into the carotid artery that feeds the dominant hemisphere not only impairs speech, but blocks self-consciousness for several tens of
seconds [14].

[‡] It may well be that subhuman primates have "bursts" of self-consciousness during
which internally recalled images are manipulated and a longer-term future is
briefly "illuminated." But there is no clear and convincing evidence that any
outcome is stored in memory for later use. In other words, it is conceivable that
some higher mammals may exhibit bursts of human-like thinking, but they seem
not to be able to do anything long-lasting with the results. There is a contentious
debate on this issue between animal psychologists and brain scientists.

"predicting" – concepts that are so important in the context of information (Table 1.1 and Fig. 3.8)! We have come across *convergence* of information in the brain, in which many different patterns relate to one, which becomes the neural signature of something. Examples are the many optical patterns elicited by the same object seen from different perspectives being mapped into one unique pattern that "represents" the object in the brain, or some specific and unique pattern elicited by, and synonymous of, the categories "a shiny red apple" or "my grandmother." We have also seen *divergence,* in which the brain can imagine and expect (predict) several alternative outcomes. Once a particular pattern has been confirmed by actual sensory input (e.g., a measurement), the corresponding change of the state of the brain (the appearance of a specific pattern and concomitant disappearance of alternative and mutually competing patterns) is the transition from "not knowing" to "knowing" (Sect. 1.6 and Sect. 3.6).

6.5 "Free Will" and the "Mind–Body Problem"

In recent years a question has arisen about "free will" (e.g., see [84]). It appears from EEG measurements of the "readiness potential" that actual brain activity related to implementing a decision comes hundreds of milliseconds, even up to a second, *before* the subject "feels the intention to act." Leaving aside the question of what it really means "to feel making a conscious decision" and doubts about the experimental determination of that critical instant of time, the key point is that in "making a decision" one given brain activity is being interrupted and replaced by another – in other words, a transition of the state of the brain takes place which would *not* have occurred without the subject having had such intention. It is a bit like a quantum system: What counts is what comes out in the end, not what we believe or feel is happening inside!

Perhaps it is advisable to use the same learning strategy in the study of brain function as in quantum mechanics: *accepting and getting used to* rather than forcing comparisons with familiar events and metaphors. A quantum system in which measurement results point to a particle following two paths at the same time while it is *not* being observed (Sect. 2.3 and Sect. 2.9) is something one has to accept and get used to, rather than trying to imagine in terms of familiar observations in the macroscopic world. In the case of the brain, viewing a sensation, an image or a thought as a horribly complex but absolutely specific spatio-temporal distribution of neural impulses is a fact one should accept and get used to, rather than trying to interpret in terms of something, inaccessible to measurement, that is "pulling the strings" from yet another level. In both quantum mechanics and brain function, what matters is how a system is prepared and how it responds – only those input and output states are amenable to measurement and verification. Their relationship embodies information as defined in Sect. 3.5: Some initial pattern (the imposed initial conditions of a quantum system, the sensory signals reaching an animal brain, or a thought generated within a human brain) leads to some output pattern (a measurement result, a behavior, or yet another thought). What happens in between is, of course, the crux of the matter. If we try to find out by poking into a quantum system with our instruments, we will destroy its state irrevocably; to connect the initial and the final states logically, we are forced to imagine, i.e., make models of its intervening behavior using an information processor (our brain) that evolved in a classical environment. Not surprisingly, we run into incompatibilities with that classical world. Trying to understand a functioning brain, the situation is different, but not much. We could, in principle, determine what each one of the $\sim 10^{11}$ neurons is doing at any time without disturbing its function (there is no quantum indeterminacy here), and we could in principle make models of equivalent self-organizing systems coherently mapping and transforming information – but this still would not provide answers understandable in terms of the linear and serial information systems with which we are familiar and – this is the key – in terms of the simplicity with which we feel our own brain is working.

I believe that a "mind–matter" or "mind–body" problem no longer exists in the neurobiological realm: From a purely scientific standpoint there is no need to consider metaphysical concepts such as "mind" or "soul," which have an existence *separate* from that of the interacting neurons that make up the living, feeling and thinking human brain. World religions need not take offense: *There is enough to marvel about the fact (call it "miracle" instead of "mere chance," call it "divine intervention" instead of "natural law") that Darwinian evolution has produced the most complex self-organizing system in the Universe as we know it – the human brain – which despite its utter complexity operates in such a highly coordinated, cooperative way that we feel ourselves as just one, able to control with a natural sense of ease and simplicity an informational machinery of unfathomable capabilities!*

References

[1] G. Aber, T. Beth, M. Horodecki, P. Horodecki, R. Horodecki, M. Rötteler, H. Weinfurter, R. Werner, A. Zeilinger (Eds.): *Quantum Information: An Introduction to Basic Theoretical Concepts and Experiments* (Springer, Berlin, Heidelberg, New York 2001)

[2] J. M. Alonso, A. N. Stepanova: Science **306**, 1513 (2004)

[3] M. A. Arbib: *Brains, Machines and Mathematics*, 2nd ed. (Springer, New York, Berlin, Heidelberg 1987)

[4] N. Arkani-Hamed, S. Dimopoulos, G. Dvali: Phys. Today **55**, 35 (2002)

[5] J. D. Bekenstein: Phys. Rev. D **7**, 2333 (1973)

[6] J. D. Bekenstein: Sci. Am. **289**, 59 (2003)

[7] C. H. Bennett: Sci. Am. **255**, 108 (1987)

[8] C. H. Bennett, in W. H. Zurek (Ed.): *Complexity, Entropy and the Physics of Information* (Addison Wesley Publishing Co. 1990) p. 137

[9] C. H. Bennett: Phys. Today **48**, 24 (1995)

[10] G. P. Berman, G. D. Doolen, R. Mainieri, V. I. Tsifrinovich: *Introduction to Quantum Computers* (World Scientific Publishing Co., Singapore 1998)

[11] D. R. Bes: *Quantum Mechanics* (Springer, Berlin, Heidelberg, New York 2004)

[12] D. Bickerton: *Language and Human Behavior* (Univ. of Washington Press, Seattle 1995)

[13] J. R. Binder, in C. T. W. Moonen, P. A. Bandettieri (Eds.): *Functional MRI* (Springer, Berlin, Heidelberg, New York 1999) p. 393

[14] H. M. Borchgrevink, in M. Clynes (Ed.): *Music, Mind and Brain* (Plenum Press, New York 1982) p. 151

[15] E. Borel: *Oeuvres*, vol. 3 (CNRS, Paris 1972 (1914))

[16] D. Bouwmeester, A. Ekert, A. Zeilinger (Eds.): *The Physics of Quantum Information* (Springer, Berlin, Heidelberg, New York 2000)

[17] J. L. Bradshaw, N. C. Nettleton: Behavioral and Brain Sci. **4**, 51 (1981)

[18] V. B. Braginski, Y. I. Vorontsov, K. S. Thorne: Science **209**, 547 (1980)

[19] V. Braitenberg: *Vehicles: Experiments in Synthetic Psychology* (MIT Press 1984)

[20] V. Braitenberg: *Das Bild der Welt im Kopf* (LIT Verlag, Münster 2004) in German

[21] J. Bricmont: Physicalia Mag. **17**, 159 (1995)

[22] L. Brillouin: *Science and Information Theory* (Academic Press Inc., New York 1963)

[23] A. Brodal: *Neurological Anatomy in Relation to Clinical Medicine* (Oxford Univ. Press 1972)

[24] E. J. Chaisson: *Cosmic Evolution* (Harvard University Press, Cambridge MA 2001)

[25] G. J. Chaitin: IBM J. of Res. and Development **21**, 350 (1977)

[26] G. P. Collins: Phys. Today **51**, 18 (1998)

[27] A. Damasio: *The Feeling of What Happens: Body and Emotion in the Making of Consciousness* (Harcourt, Inc., San Diego, New York, Boston 1999)

[28] P. Davies, in W. H. Zurek (Ed.): *Complexity, Entropy and the Physics of Information* (Addison Wesley Publishing Co. 1990) p. 61

[29] P. Davies, in J. Chela-Flores, T. Owen, F. Raulin (Eds.): *The First Steps of Life in the Universe* (Kluwer Acad. Publ., Dordrecht, The Netherlands 1990) p. 11

[30] V. Degiorgio: Am. J. Phys. **48**, 81 (1980)

[31] S. Dimopoulos, S. A. Raby, F. Wilczek: Phys. Today **44**, 25 (1991)

[32] P. A. M. Dirac: *The Principles of Quantum Mechanics* (Oxford Univ. Press 1947)

[33] R. J. Dolan: Science **298**, 1191 (2002)

[34] M. Eigen, P. Schuster: *The Hypercycle: The Principle of Natural Self-Organization* (Springer, Berlin 1979)

[35] A. Einstein, B. Podolski, N. Rosen: Phys. Rev. **47**, 777 (1935)

[36] N. J. Emery, N. S. Clayton: Science **306**, 1903 (2004)

[37] U. T. Eysel: Science **302**, 789 (2003)

[38] J. D. Fast: *Entropy* (McGraw Hill Book Co., New York 1962)

[39] R. P. Feynman: *Quantum Electrodynamics* (Benjamin, New York 1962)

[40] D. J. Freedman, M. Riesenhuber, T. Poggio, E. K. Miller: Science **291**, 312 (2001)

[41] S. J. Freeland, R. D. Knight, L. F. Landweber, L. D. Hurst: Mol. Biol. and Evolution **17**, 511 (2000)

[42] M. Gell-Mann: What is complexity?
URL www.santafe.edu/sfi/People/mgm/complexity.html

[43] M. Gell-Mann: *The Quark and the Jaguar* (W. H. Freeman, New York 1994)

[44] G. Ghirardi: Letter to readers by the editor of *Foundations of Physics* dated 7 March 2002

[45] G. Ghirardi: *Sneaking a Look at God's Cards* (Princeton Univ. Press, Princeton 2004)

[46] G. Ghirardi, A. Rimini, T. Weber: Lett. Nuovo Cim. **27**, 263 (1980)

[47] D. Giulini, E. Joos, C. Kiefer, J. Kupsch, I.-O. Stamatescu, H. D. Zeh (Eds.): *Decoherence and the Appearance of a Classical World in Quantum Theory* (Springer, Berlin, Heidelberg 1996)

[48] A. Globus, M. R. Rosenzweig, E. L. Bennett, M. C. Diamond: J. Comp. Physiol. Psych. **82**, 175 (1973)

[49] D. L. González: Med. Sci. Monit. **10**, HY11 (2004)

[50] B. Greene: *The Fabric of the Cosmos: Space, Time and the Texture of Reality* (Alfred A. Knopf, New York 2004)

[51] J. Gribbin: *Schrödinger's Kitten and the Search for Reality* (Little, Brown and Co., Boston, New York, Toronto, London 1995)

[52] H. Haken: *Information and Self-Organization: A Macroscopic Approach to Complex Systems* (Springer, Heidelberg, Berlin, New York 1988)

[53] C. J. Han, C. M. O'Tuathaigh, L. van Trigt, J. J. Quinn, M. S. Fanselau, R. Mongeau, C. Koch, D. J. Anderson: in *Proc. Natl. Acad. Sci. USA*, vol. 100 (2003) p. 13087

[54] H. F. Hartmuth: *Information Theory Applied to Space-Time Physics* (World Scientific, Singapore 1992)

[55] B. Hayes: Amer. Sci. **92**, 494 (2004)

[56] D. Hebb: *Organization and Behaviour* (Wiley and Sons, New York 1949)

[57] K. Herholz: *NeuroPET: Positron Emission Tomography in Neuroscience and Clinical Neurology* (Springer, Berlin, Heidelberg, New York 2004)

[58] G. E. Hinton: Sci. Am. **267**, 145 (1992)

[59] K.-H. Hohne: *VOXEL-MAN 3D Navigator: Brain and Skull* (Springer, Berlin, Heidelberg 2001)

[60] T. Hosokawa, D. A. Rusakov, T. V. P. Bliss, A. Fine: J. Neuroscience **15**, 5560 (1995)

[61] R. Jozsa, in H.-K. Lo, S. Popescu, T. Spiller (Eds.): *Introduction to Quantum Computation and Information* (World Scientific Publication Co., Singapore 1998) p. 49

[62] C. Koch: *The Quest for Consciousness: A Neurobiological Approach* (Roberts and Co., Englewood, Colorado 2004)

[63] T. Kohonen: *Self-Organization and Associative Memory*, 2nd ed. (Springer, Berlin, Heidelberg, New York 1988)

[64] B.-O. Küppers: *Information and the Origin of Life* (The MIT Press, Cambridge Mass. 1990)

[65] L. D. Landau, E. M. Lifshitz: *Quantum Mechanics: Non-relativistic Theory*, vol. 3: Quantum Theory (Butterworth-Heinemann, Oxford 2000) reprint

[66] D. Layzer: Sci. Am. **233**, 56 (1975)

[67] H. S. Leff, A. F. Rex (Eds.): *Maxwell's Demon: Entropy, Information, Computing* (Adam Hilger, Bristol 1990)

[68] J. Lehmann: J. Theor. Biol. **202**, 129 (2002)

[69] H.-K. Lo, S. Popescu, T. Spiller (Eds.): *Introduction to Quantum Computation and Information* (World Scientific Publication Co., Singapore 1998)

[70] E. Mach: *Science of Mechanics* (The Open Court Publ. Co., La Salle, Illinois 1942 (1893))

[71] C. von der Marlsburg, in M. Ito, Y. Miyashita, E. T. Rolls (Eds.): *Cognition, Computation and Consciousness* (Oxford Univ. Press, Oxford, New York, Tokyo 1997) p. 193

[72] D. Marr: *Vision* (W. H. Freeman and Co. 1982)

[73] W. S. McCullogh, W. A. Pitts: Bull. Math. Biophysiol. **5**, 115 (1943)

[74] J. McFadden: *Quantum Evolution* (Harper Collins, London 2000)

[75] J. R. Melcher, T. M. Talavage, M. P. Harms, in C. T. W. Moonen, P. A. Bandettieri (Eds.): *Functional MRI* (Springer, Berlin, Heidelberg, New York 1999) Chap. 32

[76] N. D. Mermin: Am. J. Phys. **71**, 23 (2003)

[77] S. B. Miller: Science **117**, 528 (1953)

[78] C. W. Misner, K. S.Thorne, J. A. Wheeler: *Gravitation* (W. H. Freeman 1973)

[79] Y. Miyashita: Science **306**, 435 (2004)

[80] J. Monod: *Chance and Necessity* (Collins, London 1972)

[81] C. T. W. Moonen, P. A. Bandettieri (Eds.): *Functional MRI* (Springer, Berlin, Heidelberg, New York 1999)

[82] M. A. Nielsen, I. L. Chuang: *Quantum Computation and Quantum Information* (Cambridge University Press 2000)

[83] M. A. Nowak, K. Sigmund: Science **303**, 793 (2004)

[84] S. S. Obhi, P. Haggard: Am. Sci. **92**, 358 (2004)

[85] R. Penrose: *The Emperor's New Mind* (Oxford Univ. Press 1989)

[86] D. Premack: Science **303**, 318 (2004)

[87] K. Pribram: *Languages of the Brain, Experimental Paradoxes and Principles in Neuropsychology* (Prentice-Hall Inc., Englewood Cliffs, New Jersey 1971)

[88] I. Prigogine: *From Being to Becoming: Time and Complexity in the Physical Sciences* (W. H. Freeman and Co. 1980)

[89] S. Rasmussen, L. Chen, D. Deamer, D. C. Krakauer, N. H. Packard, P. F. Stadler, M. A. Bedau: Science **303**, 963 (2004)

[90] J. G. Roederer: *Functions of the Human Brain: An Interdisciplinary Introduction to Neuropsychology*, Lecture Notes (Denver Res. Inst., Univ. of Denver Publ. 1976)

[91] J. G. Roederer: Found. Phys. **8**, 423 (1978)

[92] J. G. Roederer: EOS Trans. Am. Geophys. Union **62**, 569 (1981)

[93] J. G. Roederer: in R. Spintge, R. Droh (Eds.): *Music in Medicine* (Springer, Berlin, Heidelberg 1987) p. 81

[94] J. G. Roederer: *The Physics and Psychophysics of Music: An Introduction*, 3rd ed. (Springer, New York, Berlin, Heidelberg 1995)

[95] J. G. Roederer, in J. Chela-Flores, G. Lemarchand, J. Oró (Eds.):
      Astrobiology: Origins from the Big-Bang to civilization (Kluwer, Dordrecht 2000) p. 179
[96] J. G. Roederer: Entropy **5**, 3 (2003)
      URL http://www.mdpi.org/entropy/
[97] J. G. Roederer, in J. Seckbach, E. Rubin (Eds.): *The New Avenues in Bioinformatics* (Springer, Berlin, Heidelberg, New York 2004) Chap. 2
[98] J. G. Roederer, in D. Vakoch (Ed.): *Between Worlds* (The MIT Press, Boston, MA 2005) Chap. 10, in press
[99] J. E. Rose, J. F. Brugge, D. J. Anderson, J. E. Hind: J. Neurophys. **32**, 402 (1969)
[100] B. Schumacher: Phys. Rev. A **51**, 2738 (1995)
[101] Q. Shafi, in G. Fraser (Ed.): *The Particle Century* (Institute of Physics Publ. Ltd., Bristol 1998) p. 204
[102] C. E. Shannon: The Bell System Technical Journal **27**, 379 (1948)
[103] C. E. Shannon: The Bell System Technical Journal **27**, 623 (1948)
[104] C. E. Shannon, W. W. Weaver: *The Mathematical Theory of Communication* (University of Illinois Press 1949)
[105] L. S. Shulman: *Time's Arrows and Quantum Measurement* (Cambridge University Press 1997)
[106] E. M. Shuman, D. V. Madison: Science **263**, 532 (1994)
[107] E. Squires: *Conscious Mind in the Physical World* (Adam Hilger, Bristol, New York 1990)
[108] K. Staley: Science **305**, 482 (2004)
[109] M. Tegmark: Found. Phys. **9**, 25 (1996)
[110] M. Tegmark: Inf. Sciences **128**, 155 (2000)
[111] S. J. Thorpe, M. Fabre-Thorpe: Science **291**, 260 (2001)
[112] M. Tribus, E. C. McIrvine: Sci. Am. **225**, 179 (1971)
[113] S. P. Walborn, M. O. T. Cunha, S. Pádua, C. H. Monken: Am. Sci. **91**, 336 (2003)
[114] J. Z. Young: *Philosophy and the Brain* (Oxford Univ. Press, Oxford, New York 1987)
[115] H. D. Zeh, in D. Giulini, E. Joos, C. Kiefer, J. Kupsch, I.-O. Stamatescu, H. D. Zeh (Eds.): *Decoherence and the Appearance of a Classical World in Quantum Theory* (Springer, Heidelberg 1996) p. 5
[116] W. H. Zurek, in W. H. Zurek (Ed.): *Complexity, Entropy and the Physics of Information* (Addison Wesley Publishing Co. 1990) p. 73
[117] W. H. Zurek: Phys. Today **44**, 36 (1991)

Index

Printing: Krips bv, Meppel
Binding: Stürtz, Würzburg